BIOLOGY

Visualizing Life

Directed Reading Worksheets
with Answer Key

HOLT, RINEHART AND WINSTON

Harcourt Brace & Company

Austin • New York • Orlando • Atlanta • San Francisco • Boston • Dallas • Toronto • London

To the Teacher

These Directed Reading worksheets allow students to work interactively with the text, taking notes and answering questions as they read through each chapter. When the worksheets are completed, they make good tools for later study and review. One worksheet is provided for each chapter of Biology: Visualizing Life. A complete answer key for all 35 worksheets can be found in the back of this book.

Staff Credits

Editorial Development
Christopher Hess

Copyediting
Amy Daniewicz
Denise Haney
Steve Oelenberger

Prepress
Rosa Mayo Degollado

Manufacturing
Michael Roche

Acknowledgments

Writer
Bruce R. Mulkey
Science Writer
Austin, TX

Editorial Development
Pubworks

Design Development, Page Production, and Cover Design
Morgan-Cain & Associates

Cover Photography: Frans Lanting/Minden Pictures

Printed in the United States of America

ISBN 0-03-051398-7

2 3 4 5 249 00 99 98 97

Contents

Textbook Chapter

<table>
<tr><td>CHAPTER
1</td><td># The Science of Biology</td><td># Directed Reading</td></tr>
</table>

▌Section 1-1 *Biology Today*   page 5

Studying Life

Read each question, and write your answer in the space provided.

1. What is science? _____

_____

_____

2. What is the study of biology? _____

_____

_____

Biology and Medicine

Mark each statement below *T* if it is true and *F* if it is false.

3. _____ **a.** The most direct impact of biology on our lives today is in medicine.

_____ **b.** Heart disease, malaria, and cancer were the top three causes of death in the United States at the beginning of this century.

_____ **c.** Malaria is spread by the drinking of contaminated water.

_____ **d.** New strains of antibiotic-resistant bacteria that cause tuberculosis will kill about 3 million people worldwide this year.

4. List two fatal disorders that are caused by defective genes. _____

_____

_____

5. What new technique offers hope of curing fatal genetic disorders? _____

_____

_____

Biology and the Environment

Read each question, and write your answer in the space provided.

6. List three ways the exploding human population is placing stress on the environment.

_____

_____

_____

7. What actions are being taken to provide food for an expanding human population?

_____

_____

_____

8. What are the three negative effects of the waste we add to the environment? _____

_____

_____

_____

9. For what purposes are 1.3 acres of rain forest being cut or burned per second? _____

_____

_____

_____

10. What fraction of the world's species will disappear within 100 years if the destruction

of rain forests continues at the current rate? _____

_____

Biology and You

Read the question, and write your answer in the space provided.

11. List at least two questions that biology may be able to help you answer. _____

_____

_____

_____

▌Section 1-2 *Science Is a Search for Knowledge* page 11

A Case Study in Science

Read the question, and write your answer in the space provided.

1. In searching for the cause of malaria, what observations did Ross make about the location of patients who developed the disease in the field hospital? _____

_____

_____

Complete each statement by writing the correct word in each space provided.

2. The disease being discussed in this section is _____.

3. Ross proposed that malaria was spread from one patient to another by

_____ _____ .

4. Ross's testable explanation for his observations is called a(n) _____.

5. Based on his hypothesis, Ross expected several consequences. These expected consequences are called _____.

6. The parasite that causes malaria is _____ .

Read each question, and write your answer in the space provided.

7. How did Ross suspect that the parasites were transferred from a mosquito to a person?

_____

_____

8. Explain the control experiment Ross conducted with the *Anopheles* mosquitoes.

_____

_____

_____

Complete each statement by writing the correct word in each space provided.

9. In this experiment, the _____ group was not exposed to the condition suspected of causing the disease.

10. A _____ is a unifying explanation for a broad range of observations.

11. Science rejects any hypothesis not supported by _____ and the

results of _____ experiments.

Theories Have Limited Certainty

Mark each statement below *T* if it is true and *F* if it is false.

_____ **12.** There are scientific theories of which we are absolutely certain.

_____ **13.** Future evidence may cause a theory to be revised or discarded.

_____ **14.** The word "theory" is used differently by scientists and the general public.

_____ **15.** Some theories are so strongly supported that it seems unlikely they will be rejected in the future.

Read the question and write your answer in the space provided.

16. How do successful scientists study a question or problem besides by using the

scientific method? _____

_____

_____

Section 1-3 *Studying Biology*   page 16

Themes Unify Ideas

Mark each statement below *T* if it is true and *F* if it is false.

_____ **1.** A covering called a membrane surrounds a cell.

_____ **2.** The cell membrane has no control over what enters or leaves the cell.

_____ **3.** A central zone or nucleus contains the cell's genes.

_____ **4.** Proteins are frequently found in the cell's membrane.

_____ **5.** All organisms must maintain a constant internal environment in order to function properly.

_____ **6.** The human body has only one mechanism for preserving a state of constant internal conditions.

_____ **7.** Hormones are chemical signals that assist the human body in maintaining homeostasis.

_____ **8.** Nervous systems are made up of cells called electrons.

_____ **9.** In organisms composed of many cells, almost none of the cells are specialized.

_____ **10.** As the cells grow and divide, their genes manage an orderly process of change called development.

_____ **11.** The transmission of characteristics from parents to their offspring is called heredity.

_____ **12.** The mutations caused by damaged genes usually help an organism to survive.

Complete each statement by writing the correct word in each space provided.

13. The inherited change in the characteristics of organisms over time is called

_____ .

14. As a result of natural selection, organisms with favorable versions of genes are more

likely to _____ and _____ .

15. A _____ is a group of organisms that look similar and can produce fertile offspring.

Read each question, and write your answer in the space provided.

16. What is ecology? _____

_____

_____

_____

17. How may the complex web of interactions in a biological community be disrupted?

_____

_____

_____

_____

Complete each statement by writing the correct word in each space provided.

18. Living things are made up of _____ assembled into molecules.

19. The source of almost all of the energy that drives life on Earth is the

_____ .

20. The availability of _____ is a major factor in limiting the size and complexity of biological communities.

2 Discovering Life

Directed Reading

Section 2-1 *What Is Life?*

page 23

First Guesses at Defining Life

Choose the phrase from column B that best describes the term from column A.

Column A	Column B
_____ **1.** sensitivity	**a.** Small parts assembled into intricate structures
_____ **2.** death	**b.** The ability to respond to stimuli
_____ **3.** complexity	**c.** Orderly progression leading to greater specialization
_____ **4.** development	**d.** Termination of life

Life Has Five Characteristic Properties

Complete the statement by writing the correct word in each space provided.

5. The blob in Figures 2-1 through 2-3 is identified as a(n) _____

_____ _____.

Complete each statement by underlining the correct word in the brackets.

6. All living things are composed of [chlorophyll / cells].

7. They are characterized by [metabolism / complexity], because living things need energy to move, grow, and process information.

8. The process by which a living thing maintains a stable internal environment is called [heredity / homeostasis].

9. All living things [reproduce / move], and genetic information is passed from one generation to the next in the process called [sensitivity / heredity].

Section 2-2 *Basic Chemistry*

page 26

Atoms: The Basic Structural Units of Matter

Write the term that is being described in each space provided.

1. This is the smallest particle of an element that can retain its chemical properties.

_____

2. Composed of only one type of atom, over 100 types of these have been discovered.

_____

3. An atom consists of a nucleus surrounded by these tiny particles.

_____

4. The nucleus of an atom contains these two types of particles: _____

and _____.

5. These are regions of space outside the nucleus where electrons travel.

_____

Formation of Bonds Stabilizes Atoms

Mark each statement below _T_ if it is true and _F_ if it is false.

_____ **6.** An atom containing fewer electrons than protons has a negative ionic charge.

_____ **7.** A partially filled outer energy level makes an atom stable.

_____ **8.** Atoms gain or lose electrons to fill their outer energy levels and form ions.

_____ **9.** Sodium chloride is an example of an ionic compound.

_____ **10.** An ionic bond is the force holding molecules together.

_____ **11.** Water consists of an atom of oxygen that has formed a covalent bond with two atoms of hydrogen.

Explain the difference between each of the following sets of terms in the space provided.

12. ionic bond, covalent bond _____

_____

_____

_____

13. molecule, element _____

_____

_____

_____

14. single bond, double bond, triple bond _____

_____

_____

▮Section 2-3 *Molecules of Life*   page 29

Organic Molecules Are Derived From Carbon

Complete each statement by writing the correct word in each space provided.

1. The atom that is most closely associated with living things is _____ .

2. Organic molecules are linked by _____ bonds.

3. _____ , _____ , _____ , and

_____ make up the four major classes of macromolecules.

4. When a bond that holds the atoms in a molecule together is formed or broken, a

_____ _____ has taken place.

Carbohydrates Are Energy Sources

Place a check mark next to each accurate statement.

_____ **5.** Carbohydrates are good energy sources because their bonds store a large amount of energy.

_____ **6.** If a carbohydrate molecule contained six carbon atoms, it would also have six hydrogen atoms.

_____ **7.** Polysaccharides are insoluble in water.

_____ **8.** Animals, including humans, store glucose in the form amylose.

_____ **9.** Humans lack the microorganisms in their digestive system to break down cellulose.

Lipids Store Energy

Complete each statement by underlining the correct word or phrase in the brackets.

10. Lipids will dissolve in [water / oil].

11. Fats are composed of three fatty acid molecules joined to a molecule of [glycerol / protein].

12. At room temperature, [saturated / unsaturated] fats are usually liquids.

13. People whose diets are high in saturated fats are [more likely / less likely] to suffer from heart disease.

Proteins Provide Structure and Increase Reaction Rate

Read each question, and write your answer in the space provided.

14. What is a protein? _____

_____

_____

15. Explain how a protein's amino acid sequence determines its shape. _____

_____

_____

_____

16. What roles do proteins play in organisms? _____

_____

_____

_____

Nucleic Acids Contain Genetic Information

Choose the phrase from column B that best describes the term from column A.

Column A	Column B
_____ **17.** nucleic acid	**a.** Subunits of DNA and RNA
_____ **18.** nucleotides	**b.** Serve as scaffolds in the assembly of proteins in the cell
_____ **19.** DNA	**c.** Protein-DNA bundle found in the cells of the human body
_____ **20.** RNA	**d.** One of the major classes of organic macromolecules
_____ **21.** chromosomes	**e.** Its structure is known as a double-helix

3 Cells

CHAPTER

Directed Reading

■ Section 3-1 *At the Boundary of the Cell*   page 41

Cells: The Smallest Vessels of Life

Write the term that is being described in each space provided.

1. The smallest unit that can carry out all of the activities necessary for life. _____

2. This regulates what goes into and out of a cell. _____

3. All of the living things in Figure 3-1 are made up of these tiny compartments.

_____

What Limits the Size of a Cell?

Mark each statement below *T* if it is true and *F* if it is false.

_____ **4.** Most cells are big enough to be seen with the naked eye.

_____ **5.** Cells work efficiently because their interior structures are near the cell membrane.

_____ **6.** The size of a cell is limited by the relationship between its surface area and its volume.

Water and the Cell

Complete each statement by writing the correct word in each space provided.

7. All cells are surrounded by _____ .

8. A water molecule is made of two hydrogen atoms and one _____ atom.

9. A water molecule is a _____ molecule because it has a partial negative charge on one side and partial positive charge on the other side.

10. When the atom of a water molecule is attracted to the oxygen atom of another water

molecule, a _____ bond is formed.

Water and the Cell Membrane

Read each question, and write your answer in the space provided.

11. Why do water and oil separate when they are mixed together? _____

_____

_____

_____

12. Why is a lipid considered a nonpolar molecule? _____

_____

_____

_____

■ Section 3-2 *Membrane Architecture* page 46

Structure of the Lipid Bilayer

Complete each statement by writing the correct word in each space provided.

1. The framework of the cell _____ is formed by phospholipid molecules.

2. The _____ of a phospholipid molecule is polar, and the long

_____ are nonpolar.

3. The _____ bilayer is made up of a double layer of phospholipids.

Characteristics of the Lipid Bilayer

Read each question, and write your answer in the space provided.

4. What are the two most important characteristics of the lipid bilayers found in cell

membranes? _____

_____

_____

5. How do most food molecules and other substances enter or leave a cell? _____

_____

_____

Roles of Cell Membrane Proteins

Complete each statement by underlining the correct word in the brackets.

6. [Proteins / Phospholipids] that stick out of the cell membrane may serve as channels, receptors, or markers.

7. The passageway shown in Figure 3-8 is called a [receptor / channel].

8. A [receptor / marker] protein can pass information from outside the cell membrane into the cell.

9. [Channels / Markers] serve as "name tags" on every cell.

Protein Structure and the Cell Membrane

Mark each statement below *T* if it is true and *F* if it is false.

_____ **10.** All proteins are polar.

_____ **11.** The two end sections of a protein in a cell membrane are polar and form hydrogen bonds with water.

_____ **12.** The middle section of a protein such as the one shown in Figure 3-11 is made up of amino acids that are primarily nonpolar.

■ Section 3-3 *Inside the Cell*   page 50

Two Types of Cells

Complete each statement by writing the correct word in each space provided.

1. All cells fit into one of two categories: _____ or _____ .

2. The _____ of a eukaryotic cell contains DNA within chromosomes.

3. _____ are the only living prokaryotes.

4. Both eukaryotic and prokaryotic cells have _____ ,

_____ , and _____ .

Eukaryotic Cells Have Compartments

Place a check mark next to each accurate statement.

_____ **5.** The cytoplasm of a eukaryotic cell consists of the cytosol and organelles.

_____ **6.** Each organelle in a cell carries out the same function.

_____ **7.** Muscle cells, bone cells, nerve cells, and sperm cells are all examples of specialized prokaryotic cells.

_____ **8.** Unicellular eukaryotes were able to evolve into multicellular organisms because of specialization.

Organelles: A Cell's Laborers

Choose the phrase from column B that best describes the term from column A.

Column A	Column B
_____ **9.** nucleus	**a.** Convert food into energy in plant and animal cells
_____ **10.** mitochondria	**b.** The cell's packaging and distribution center
_____ **11.** chloroplasts	**c.** Act as a highway system through the cytoplasm
_____ **12.** endoplasmic reticulum	**d.** Storage center for the cell's DNA
_____ **13.** Golgi apparatus	**e.** Use the process of photosynthesis to make food

Plant Cells Differ From Animal Cells

Fill in the table by checking *Plants* if the structure is found in plants, or *Animals* if the structure is found in animals. Some structures are found in both.

14.

Structure	Plants	Animals
a. cell membrane		
b. ribosomes		
c. nucleus		
d. cell wall		
e. mitochondria		
f. chloroplasts		
g. large central vacuole		

The Origin of Eukaryotic Cells

Read the question and write your answer in the space provided.

15. List at least three pieces of evidence that support the theory of endosymbiosis. _____

_____

_____

_____

_____

Viewing the Cell

Mark each statement below *T* if it is true and *F* if it is false.

_____ **16.** If a compound microscope has an eyepiece lens that magnifies 10 times and an objective lens that magnifies 50 times, the specimen will appear to be 500 times larger than it actually is.

_____ **17.** It is possible to view living specimens with a transmission electron microscope.

_____ **18.** A scanning electron microscope forms an image of specimens that can be viewed on a video screen.

_____ **19.** A disadvantage of the scanning tunneling microscope is that it cannot be used to view living cells.

CHAPTER

4 The Living Cell

Directed Reading

Section 4-1 *How Cells Communicate*

page 63

Cell Communication in Multicellular Organisms

Mark each statement below *T* if it is true and *F* if it is false.

_____ **1.** The methods of communication in the cells of your body and the cells of bacteria are essentially the same.

_____ **2.** Multicellular organisms could not exist without communication among their cells.

Receptors Transmit Information

Complete the statement by writing the correct word in each space provided.

3. Some _____ _____ are located on the cell's surface and transmit information from outside the cell to its interior.

4. Most receptor proteins extend through the cell's _____ .

5. Protein _____ are chemical signals that must bind to a cell's surface receptor proteins to deliver their messages.

6. One example of a protein hormone is _____ .

7. _____ , _____ , and _____ are hormones that deliver their message by binding to receptor proteins inside the cell.

Some Channels Can Respond to Electricity

Complete each statement by underlining the correct word or phrase in the brackets.

8. The human nervous system is composed of cells that can respond to [electrical / chemical] signals transported by ions.

9. Nervous system cells have proteins called [hormone-sensitive / voltage-sensitive] channels embedded in the cell membrane.

10. When ions carrying an electrical signal arrive at a channel on the cell membrane of a neuron, the channel gate [closes / opens].

Markers Identify Cells

Write the term that is being described in each space provided.

11. These cell membrane proteins differentiate your cells from other cells.

_____ _____ _____

12. Cell surface markers are so distinctive that exact matches occur only between these

relatives. _____ _____

■Section 4-2 *Movement of Substances Into and Out of Cells* page ●

Diffusion and Osmosis

Choose the phrase from column B that best describes the term from column A.

Column A	Column B
1. diffusion	**a.** The force created when water causes a cell to swell.
2. osmosis	**b.** The mixing of two substances by the random motion of molecules
3. osmotic pressure	**c.** The process that takes place when water molecules pass through the gaps between phospholipids in the cell membrane.

Channels and Pumps Provide Selective Transport

Read each question and write your answer in the space provided.

4. Briefly describe the two kinds of selective transport. _____

_____

_____

_____

5. How do sodium-potassium pumps and proton pumps support the efficient functioning

of your body? _____

_____

_____

_____

6. What causes cystic fibrosis? _____

_____

_____

_____

Moving Large Particles

Explain the difference in the meanings of the terms in the following pair:

7. endocytosis / exocytosis _____

_____

_____

_____

5 Energy and Life

CHAPTER

Directed Reading

▌Section 5-1 *Cells and Chemistry*

page 77

Chemical Reactions in Living Things

Choose the phrase from column B that best describes the term from column A.

Column A	Column B
_____ **1.** reactants	**a.** The energy needed to start a chemical reaction
_____ **2.** products	**b.** Proteins that can speed up a chemical reaction
_____ **3.** metabolism	**c.** Starting materials in a chemical reaction
_____ **4.** activation energy	**d.** Ending materials in a chemical reaction
_____ **5.** enzymes	**e.** Substance that increases the speed of a chemical reaction without being used up
_____ **6.** catalyst	**f.** All of the chemical reactions taking place in an organism

Actions of Biological Catalysts

Complete each statement by writing the correct word in each space provided.

7. Without the enzyme _____ _____, a human's blood would quickly be poisoned with carbon dioxide.

8. The molecule on which an enzyme acts is called a(n) _____.

9. When an enzyme's substrate fits into the enzyme's _____

_____ , the activation energy required for a chemical reaction to take place is lowered.

10. When the _____ in your body fluids is too high or too low, most enzymes cannot function properly.

▌Section 5-2 *Cells and Energy*

page 82

How Cells Use Energy

Mark each statement below *T* if it is true and *F* if it is false.

_____ **1.** The white blood cell in Figure 5-7 is using energy to change its shape.

_____ **2.** The reaction that takes place during the burning of food in living cells is the same as the one that occurs during the burning of logs in a campfire.

_____ 3. ADP supplies the energy that a cell needs to build new molecules.

_____ 4. A reaction that is driven by the energy released when ATP is converted to ADP is called a coupled reaction.

Energy Flow in the Living World

Write the term that is being described in each space provided.

5. The process that takes place when light energy from the sun is converted to

   carbohydrates and other organic molecules by plants. _____

6. The process used by all living organisms to obtain energy from carbohydrates and

   other organic molecules. _____

▌Section 5-3 *Photosynthesis*   page 85

Harnessing the Sun's Energy

Place a check mark next to each accurate statement.

_____ 1. About one percent of the light energy that reaches the Earth from the sun is converted to chemical energy through photosynthesis.

_____ 2. The carbon dioxide gas in our atmosphere is the result of the release of huge amounts of this gas during photosynthesis.

Stage 1: Capturing Light Energy

Complete each statement by writing the correct word in each space provided.

3. Humans can see only _____ light in the electromagnetic spectrum.

4. _____ is the pigment that enables you to see certain wavelengths of light.

5. The primary pigment in plants, _____ , absorbs red and blue light, causing plants to look green.

6. Electrons within certain chlorophyll molecules are elevated to a higher energy level

   when sunlight strikes a(n) _____ .

7. During photosynthesis, plants attain electrons by splitting _____ molecules.

Stage 2: Using Light Energy to Make ATP and NADPH

Study the following steps of photosynthesis to determine the order in which they take place. Write the number of each step in the blank provided.

The correct order of events during photosynthesis is:

_____ **8.** An electron becomes excited by the light.

_____ **9.** Protons inside the thylakoid are driven by diffusion through a protein channel.

_____ **10.** Light strikes a chlorophyll molecule in the membrane of a thylakoid.

_____ **11.** The energy of the electron powers the pumping of protons across the thylakoid membrane and into the interior of the thylakoid.

_____ **12.** The force of the protons leaving the thylakoid adds a phosphate to ADP, forming ATP.

Stage 3: Building Carbohydrates

Read each question and write your answer in the space provided.

13. Write the equation for the overall process of photosynthesis. _____

_____

14. Summarize what takes place in the Calvin cycle. _____

_____

_____

_____

_____

_____

■ Section 5-4 *Cellular Respiration*   page 90

Releasing Energy From Organic Molecules

Complete each statement by underlining the correct word or phrase in the brackets.

 1. The first stage of cellular respiration is called [glycolysis / photosynthesis].

 2. [Glycolysis / Oxidative respiration] is the second stage of cellular respiration in most living things.

Stage 1: Glycolysis

Complete each statement by writing the correct word in each space provided.

3. During gylcolysis, one molecule of glucose is split into two molecules of

_____ _____.

4. The process of glycolysis is followed by another set of reactions: either

_____ or _____ _____.

5. During fermentation, some organisms break down pyruvic acid into carbon dioxide

and _____ _____.

6. When your blood circulation cannot remove excess_____

_____ quickly enough from your muscles during exercise, it builds
up and your muscles become tired.

Stage 2: Oxidative Respiration

Mark each statement below *T* if it is true and *F* if it is false.

_____ **7.** The complete breakdown of a glucose molecule during oxidative respiration
requires six oxygen molecules.

_____ **8.** Oxidative respiration occurs in just one step: the Krebs cycle.

_____ **9.** Two carbon dioxide molecules, one ATP molecule, and additional NADH molecules are produced during the Krebs cycle.

_____ **10.** The electron transport chain is composed of a series of carbohydrates in the
membranes within mitochondria.

_____ **11.** The electron transport chain would become clogged with spent electrons if no
oxygen were available to accept these electrons.

Regulating Cellular Respiration

Read the question and write your answer in the space provided.

12. Describe the process that regulates cellular respiration. _____

_____

_____

_____

6 Cell Reproduction

Directed Reading

▌Section 6-1 *Chromosomes*

page 101

Chromosome Structure

Complete each statement by writing the correct word in each space provided.

1. The strand of DNA and protein shown in Figure 6-1 is called a(n) _____ .

2. Eukaryotic chromosomes are made of _____ , a blend of DNA and protein.

3. The strand of DNA from a chromosome could be as long as _____ cm if it were laid out in a straight line.

Chromosome Number

Mark each statement below *T* if it is true and *F* if it is false.

_____ **4.** A karyotype of almost any cell in your body would show 46 chromosomes arranged into 23 pairs.

_____ **5.** Human cells that contain 23 pairs of chromosomes are said to be haploid.

_____ **6.** Sperm cells are said to be haploid.

_____ **7.** Sex chromosomes designate whether a human is a male or a female.

_____ **8.** Females have two X chromosomes, while males have two Y chromosomes.

▌Section 6-2 *Mitosis and Cell Division*

page 104

Cell Reproduction Differs in Bacteria and Eukaryotes

Write the term that is being described in each space provided.

1. When a cell divides, each resulting new cell must contain a complete copy of this to be healthy. _____

2. During reproduction, this type of cell splits into two distinct cells, each with its own circle of DNA. _____

How Eukaryotic Nuclei Divide: Mitosis

Read each question and write your answer in the space provided.

3. In terms of their DNA, how does eukaryotic cell division differ from cell division in a

prokaryotic cell? _____

_____

_____

4. List the four phases of mitosis in order. _____

_____

_____

_____

5. What is the end result of mitosis and cytokinesis? _____

_____

_____

_____

The Cell Cycle

Study the following phases of the cell cycle to determine the order in which they take place. Write the number of each step in the blank provided.

_____ **6.** G2 phase: Growth and preparation for nuclear division

_____ **7.** Cytokinesis: Cytoplasm divides

_____ **8.** G1 phase: Cell growth

_____ **9.** Mitosis: Nucleus divides into two nuclei

_____ **10.** S phase: DNA copied

Controlling the Cell Cycle

Complete each statement by underlining the correct word or phrase in the brackets.

11. Animal cells use a class of proteins called [cyclins / RNA] to help control the phases of the cell cycle.

12. If cells grow and divide without restraint, a [gamete / tumor] may result.

13. Cancer results from a malfunction of [meiosis / the cell cycle].

14. Exposure to radiation and the use of tobacco products are suspected of contributing to the development of [cancer / tuberculosis].

▌Section 6-3 *How Gametes Form: Meiosis* page 110

Making Haploid Cells

Complete each statement by writing the correct word in each space provided.

1. _____ are haploid reproductive cells.

2. Gametes are formed during a two-stage form of nuclear division called

_____.

3. Meiosis produces four haploid cells, each with _____ as much genetic material as the original cell.

The Importance of Crossing-Over

Place a check mark next to each accurate statement.

_____ **4.** A diploid cell is formed when two gametes unite in fertilization.

_____ **5.** The phenomenon taking place in Figure 6-8 is called crossing-over.

_____ **6.** At the conclusion of crossing-over, the two sets of chromosomes are still identical.

_____ **7.** In terms of evolution, crossing-over is relatively unimportant.

7 Genetics and Inheritance

CHAPTER

Directed Reading

▮Section 7-1 *The Work of Gregor Mendel* page 117

Gregor Mendel and the Garden Pea

Mark each statement below *T* if it is true and *F* if it is false.

_____ **1.** The scientific study of heredity is called genetics.

_____ **2.** Gregor Mendel studied patterns of heredity by performing experiments on oak trees.

_____ **3.** Mendel was among the first to use a mathematical approach to the study of heredity.

_____ **4.** The plant shown in Figure 7-1 worked well in genetic studies because the female and male reproductive parts are found in different parts of the organism.

Mendel's Experiments

Complete each statement by writing the correct word in each space provided.

5. In Step 1 of Mendel's experiments, the white-flowering strain of pea plants produced

only _____ flowers.

6. In Step 2 of Mendel's experiments, he found only _____ flowers.

7. In Step 3 of Mendel's experiments, plants with purple flowers outnumbered the plants

with white flowers by a _____ ratio.

Mendel Developed a Model to Explain His Results

Choose the term from column B that best matches the phrase in column A, and write your answer in the space provided.

	Column A	Column B
_____	**8.** Genes that exist in more than one form	**a.** alleles
_____	**9.** Individual having two copies of different alleles	**b.** homozygous
_____	**10.** The many alleles that an organism has	**c.** heterozygous
_____	**11.** Individual having two copies of the same allele	**d.** dominant
_____	**12.** An organism's physical appearance	**e.** recessive
_____	**13.** The allele that is expressed	**f.** genotype
_____	**14.** The allele that is not expressed	**g.** phenotype

Visualizing Mendel's Model

Write the term that is being described in each space provided.

15. A model that shows all of the possible outcomes of a cross. _____

16. *WW* or *Ww* indicates the _____ of the purple flower in Figure 7-3.

17. The likelihood that something will happen. _____

Read each question and write your answer in the space provided.

18. Name the genotype(s) that result(s) from the cross shown in Figure 7-4. _____

19. What is the ratio of purple-flowering plants to white-flowering plants in the offspring

of the cross shown in Figure 7-4? _____

Mendel's Conclusions

Complete each statement by writing the correct word in each space provided.

20. The law of _____ states Mendel's conclusion that two factors for a trait separate into different gametes.

21. The law of _____ _____ states Mendel's conclusion that two or more pairs of alleles segregate independently of one another during gamete formation.

▌Section 7-2 *Patterns of Inheritance* page 123

Predicting the Outcome of a Cross

Read each question and write your answer in the space provided.

1. What type of cross is shown in Figure 7-5? _____

2. How many different phenotypes are possible in the offspring in the cross shown in

Figure 7-5? How many genotypes? _____

_____

_____

_____

Visualizing a Dihybrid Cross

Read each question and write your answer in the space provided.

3. What traits do the alleles *Y, y, R,* and *r* symbolize in the dihybrid cross shown in

Figure 7-6? _____

_____

_____

4. What phenotypes will be found among the offspring of the cross shown in Figure 7-6?

In what ratio? _____

_____

_____

Other Factors That Influence Heredity

Complete each statement by underlining the correct word or phrase in the brackets.

5. When red-flowering snapdragons are crossed with white-flowering snapdragons, the
resulting pink flowers of the offspring provide an example of [complete dominance /
incomplete dominance].

6. The roan coat of the horse in Figure 7-8 is a result of [incomplete dominance /
codominance].

7. The possible genotypes for the human blood group A are [$I_B I_B$ or $I_B i$ / $I_A I_A$ or $I_A i$].

8. Human height, weight, and hair and skin color are all [polygenic / environmental] traits.

■ Section 7-3 *Human Genetic Disorders* page 127

Mutations Are Changes in Genes

Place a check mark next to each accurate statement in the space provided.

_____ **1.** A mutation results when genes are damaged or copied incorrectly.

_____ **2.** Most mutations improve an individual.

_____ **3.** Typically, mutations are recessive and not expressed.

Genetic Disorders

Complete each statement by writing the correct word in each space provided.

4. _____ _____ is caused by a mutation in a gene that codes for a protein responsible for transporting chloride ions.

5. People who are heterozygous for the sickle cell gene are more resistant to the deadly tropical disease _____ .

6. Look at the cross shown in Figure 7-15. Of the offspring in this cross who inherit the gene that causes hemophilia, only the _____ will develop this genetic disorder.

7. People with _____ _____ have three copies of chromosome 21.

Genetic Counseling and Gene Therapy

Read the question and write your answer in the space provided.

8. Explain why a recently married couple might wish to have a family pedigree prepared.

_____

_____

_____

Finding Cures for Genetic Disorders

Complete the statement by underlining the correct word or phrase in the brackets.

9. [Phenylketonuria / Sickle cell anemia], a recessive disorder that can cause mental retardation, can be avoided if it is diagnosed shortly after birth.

Complete the statement by writing the correct word in each space provided.

10. Gene technology is making it possible to correct genetic disorders by replacing copies of _____ genes with copies of _____ genes.

<table>
<tr><td>CHAPTER</td><td># 8</td><td>How Genes Work</td><td>Directed Reading</td></tr>
</table>

Section 8-1 *Understanding DNA* page 137

How Scientists Identified the Genetic Material

Complete each statement by writing the correct word in each space provided.

1. Griffith found that one strain of the bacterium *Pneumococcus* was _____ and killed mice.

2. In Griffith's experiment, nonvirulent bacteria had changed to virulent bacteria through

the process of _____.

3. Avery and his colleagues proved that _____ was the genetic material of the bacteria in the experiment.

4. When _____ and _____ examined the remains of bacteria infected by the 32P-labeled viruses, they found DNA inside the bacteria and in the new viruses that resulted from the infection.

How Scientists Determined the Structure of DNA

Choose the phrase from column B that best describes the term from column A, and write your answer in the space provided.

Column A	Column B
_____ **5.** nucleotides	**a.** Adenine and guanine belong to this class of molecules.
_____ **6.** purines	**b.** A subunit of DNA that is composed of deoxyribose, a phosphate group, and a base
_____ **7.** pyrimidines	
_____ **8.** double helix	**c.** The structure of a DNA molecule
	d. Cytosine and thymine belong to this class of molecules.

How DNA Is Copied

Mark each statement below *T* if it is true and *F* if it is false.

_____ **9.** The process of copying RNA prior to cell division is called replication.

_____ **10.** If the sequence of one strand of a DNA molecule is ATTGCAT, the sequence of the partner strand must be TAACGTA.

_____ **11.** Replication does not maintain the sequence of bases in an organism's DNA.

_____ **12.** Mutagens may cause mutations that alter the structure of DNA.

▪ Section 8-2 *How Proteins Are Made*   page 141

The Transfer of Genetic Information

Write the term that is being described in each space provided.

1. DNA is used as a blueprint to make this molecule. _____

2. The use of genetic information in DNA to make proteins _____

3. The stage of gene expression during which an RNA copy of a gene is made

_____

4. The stage of gene expression during which different kinds of RNA assemble amino

acids into a protein molecule _____

How DNA Makes RNA

Complete each statement by underlining the correct word in the brackets.

5. Human cells protect genetic information by keeping the [DNA / RNA] in the nucleus.

6. One of the differences in the structure of DNA and RNA is that the base thymine is

replaced by the base [guanine / uracil] in RNA.

7. During [transcription / translation], a portion of the double helix unwinds and sepa-

rates a section of the two DNA strands.

8. As RNA polymerase moves along one strand of the DNA, the enzyme joins

nucleotides into a complementary chain of single-stranded [DNA / RNA].

The Genetic Code

Complete each statement by writing the correct word in each space provided.

9. Each nucleotide triplet in mRNA is called a(n) _____ .

10. Scientists found that the codon _____ codes for the amino acid

phenylalanine.

11. There are _____ different possible three-letter codons in the

genetic code.

12. It is believed that the genetic code had originated by the time

_____ evolved.

How RNA Makes Proteins

Study the following phases of translation to determine the order in which the steps take place. Write the number of each step in the blank provided.

_____ **13.** tRNA carries amino acids to the ribosome according to the three-base codon.

_____ **14.** As the mRNA passes through the ribosome, the amino acids delivered by tRNA are added to the end of the growing protein chain.

_____ **15.** When the end of the mRNA sequence is reached, the newly made protein is released into the cell.

_____ **16.** mRNA enters the cytoplasm and binds to a ribosome.

▌Section 8-3 *Regulating Gene Expression*   page 146

Switching Genes On and Off

Place a check mark next to each accurate statement.

_____ **1.** A repressor protein blocks transcription by preventing the RNA polymerase from moving along the gene to a binding site.

_____ **2.** When an inducer binds to a repressor protein, the repressor protein binds more tightly to the DNA.

_____ **3.** Many genes need the assistance of a regulatory protein called an activator to partially unwind in order to expose the bases at the regulatory site.

_____ **4.** Bacteria use transcription factors called enhancers to control the onset of transcription.

Architecture of the Gene

Read each question and write your answer in the space provided.

5. Explain the difference between exons and introns. _____

_____

_____

6. Describe the discovery made by the geneticist Barbara McClintock. _____

_____

_____

_____

9 Gene Technology

CHAPTER

Directed Reading

■ Section 9-1 *The Revolution in Genetics*

page 153

What Is Genetic Engineering?

Complete each statement by underlining the correct word in the brackets.

1. Cohen and Boyer revolutionized genetics by producing recombinant [DNA / RNA].

2. In this 1973 experiment, genetically engineered [bacteria / human] cells produced frog rRNA.

3. Moving genes from the chromosomes of one organism to those of another is called [genetic / chemical] engineering.

How to Move a Gene From One Organism to Another

Mark each statement below *T* if it is true and *F* if it is false.

_____ 4. In a genetic engineering experiment, scientists must first cut the source chromosome into fragments so that they can isolate the gene that is to be transferred.

_____ 5. Cohen and Boyer used the restriction enzyme CTTAAG to cut the frog DNA.

_____ 6. The frog DNA fragments and the cut bacteria plasmids in Cohen and Boyer's experiment joined together in circular DNA molecules because they had complementary sticky ends.

_____ 7. Vectors can never be used as delivery agents in genetic engineering experiments.

_____ 8. As the bacteria in Cohen and Boyer's experiment reproduced, they produced clones of themselves.

_____ 9. In this experiment, the antibiotic tetracycline killed all of the bacterial cells.

Safety Considerations

Read the question and write your answer in the space provided.

10. List two specific concerns about the use of genetic engineering technology.

_____

_____

_____

Section 9-2 *Transforming Agriculture*   page 158

Getting Genes Into Eukaryotic Cells

Place a check mark next to each accurate statement.

_____ **1.** When scientists wish to use the Ti plasmid to transfer genes into plant cells, they must first remove the tumor-causing gene from the plasmid.

_____ **2.** The Ti plasmid is useful in inserting DNA into corn, rice, and wheat plants.

_____ **3.** DNA particle guns have been used to insert genes into the DNA of chloroplasts.

Resistance to Herbicides

Read each question and write your answer in the space provided.

4. Why are crop plants that are resistant to glyphosate so significant to farmers?

_____

_____

Nitrogen Fixation

Read each question and write your answer in the space provided.

5. Briefly explain why nitrogen fixation is essential to plants and how this process works.

_____

_____

_____

Resistance to Insects

Complete each statement by writing the correct word in each space provided.

6. More than 40 percent of all chemical pesticides are used to stamp out insects that feed

on _____ plants.

7. Biologists have made _____ plants resistant to hornworms by inserting the genes for specific bacterial enzymes into the plants.

Genetic Engineering in Livestock

Mark each statement below *T* if it is true and *F* if it is false.

_____ **8.** Bovine growth hormone is injected into dairy cows to decrease milk production.

_____ **9.** Genetic engineering is now used to make bovine growth hormone in the laboratory.

_____ **10.** Leaner and faster-growing cattle and hogs are likely to result from genetic engineering.

Section 9-3 *Advances in Medicine* page 162

Making Miracle Drugs

Write the term that is being described in each space provided.

1. This disease affects a person whose body cannot produce insulin. _____

2. This drug can now be produced inexpensively thanks to the procedures shown in

Figure 9-11. _____

Making Vaccines

Study the following steps required to make a vaccine against genital herpes. Determine the order in which the steps take place, and write the number of each step in the blank provided.

_____ **3.** The DNA fragment is inserted into a cowpox virus, and the virus makes herpes surface proteins.

_____ **4.** A person is injected with the genetically engineered virus, stimulating the person's body to make antibodies against the genital herpes virus.

_____ **5.** A DNA fragment that contains the gene that codes for the herpes surface protein is obtained.

_____ **6.** The virus that causes genital herpes is obtained from the cells of a person who is infected with the disease.

Human Gene Therapy

Read each question and write your answer in the space provided.

7. Why has gene therapy been unsuccessful in curing human genetic diseases in the past?

_____

_____

_____

8. Briefly describe the first attempt at human gene therapy. _____

_____

_____

_____

9. How have genetic engineers increased the effectiveness of TNF? _____

_____

_____

_____

Identifying Sequences in Genes

Choose the phrase from column B that best describes the term from column A.

Column A	Column B
_____ **10.** DNA fingerprint	**a.** A process that can make millions of copies of a gene in a short time
_____ **11.** PCR	**b.** The effort to determine the nucleotide base sequence of every human gene
_____ **12.** human genome	**c.** The pattern of short black bands on a piece of film produced from a DNA sample
_____ **13.** Human Genome Project	**d.** The entire collection of genes within human cells

10 Evolution and Natural Selection

CHAPTER

Directed Reading

▮ Section 10-1 *Charles Darwin*

page 173

Voyage of the *Beagle*

Mark each statement below *T* if it is true and *F* if it is false.

_____ **1.** Figure 10-1 shows Charles Darwin's five-year voyage around the world on the H.M.S. *Beagle.*

_____ **2.** Darwin's father sent him to Cambridge University for medical training.

_____ **3.** During his journey, Darwin began to doubt that species did not change once they were created.

Darwin's Finches

Complete each statement by writing the correct word in each space provided.

4. During his visit to the _____ Islands, Darwin found that the plants and animals there closely resembled those of the nearby coast of South America.

5. The change in species over time is referred to as _____ .

Darwin's Mechanism for Evolution

Complete each statement by underlining the correct word in the brackets.

6. The publishing of [Darwin's / Wallace's] book *On the Origin of Species by Means of Natural Selection* caused a great controversy.

7. Natural selection occurs because organisms with traits that enable them to survive are [less / more] likely to reproduce.

▮ Section 10-2 *The Evidence for Evolution*

page 176

Understanding the Fossil Record

Write the term that is being described in each space provided.

1. Traces of dead organisms _____

2. For a fossil to form, the dead organism must be covered by this substance.

_____

How Fossils Are Dated

Mark each statement below *T* if it is true and *F* if it is false.

_____ 3. Carbon-14 is typically used to date fossils because it has a longer half-life than most other isotopes.

_____ 4. Scientists have used radioactive dating to determine that the Earth is about 4.5 billion years old.

_____ 5. Darwin predicted that there would be no intermediate stages between earlier and more recent species.

_____ 6. Because they are descendants of four-legged land animals, modern whales still have a pelvis but no rear legs.

Comparing Organisms

Choose the example from column B that best fits the method for comparing organisms in column A, and write your answer in the space provided.

Column A	Column B
_____ 7. homologous structures	**a.** A whale's pelvis
_____ 8. vestigial structures	**b.** A human embryo and a chicken embryo
_____ 9. developmental patterns	**c.** The cytochrome found in human and chimpanzee amino acids
_____ 10. DNA or protein sequences	**d.** The bones in a human arm, a bird wing, or the front fin of a dolphin

■ Section 10-3 *Natural Selection* page 180

How Natural Selection Causes Evolution

Study the following steps in the process of natural selection to determine the order in which the steps take place. Write the number of each step in the blank provided.

_____ 1. Natural selection causes genetic change.

_____ 2. Variation is the raw material for natural selection.

_____ 3. Species adapt to their environment.

_____ 4. Only some individuals survive and reproduce.

_____ 5. Living things face a constant struggle for existence.

The Peppered Moth: Natural Selection in Action

Explain what Kettlewell learned about the survival rate of the light and dark moths in each of the three steps of his experiments in the spaces below.

6. Step 1 _____

_____

_____

_____

_____

7. Step 2 _____

_____

_____

_____

_____

8. Step 3 _____

_____

_____

_____

The Puzzle of Sickle Cell Anemia

Complete each statement by writing the correct word in each space provided.

9. Figure 10-10 shows the red blood cells of people who have _____

_____ _____ .

10. People who are heterozygous for the sickle cell allele are resistant to the disease of

_____ .

11. Because natural selection is acting on the sickle cell allele in opposite directions in

central Africa, _____ selection is taking place.

12. Sickle cell anemia is much less common in the United States because of

_____ selection.

How Species Form

Place a check mark next to each accurate statement.

_____ **13.** The sparrows shown in Figure 10-11 are members of the same ecological race.

_____ **14.** Ecological races are members of the same species that differ genetically.

_____ **15.** Ecological races may become distinct species over time due to divergence.

Does Evolution Occur in Spurts?

Read the question and write your answer in the space provided.

16. Briefly explain how the hypothesis of gradualism differs from the punctuated

equilibria hypothesis. _____

_____

_____

_____

_____

_____

11 History of Life on Earth | Directed Reading

▮ Section 11-1: *Origin of Life* page 191

How Did Life Begin?

Complete each statement by writing the correct word in each space provided.

1. The _____ explanation for the existence of life on Earth maintains that life arrived here from a planet outside our solar system.

2. Those who conclude that divine forces put life on Earth believe

in _____ .

3. Most scientists now believe that life arose from _____ matter.

Origin of Life's Chemicals

Choose the statement from column B that best matches the ideas of the scientists listed in column A, then write your answer in the space provided.

Column A	Column B
_____ **4.** Oparin and Haldane	**a.** discovered that RNA molecules can catalyze chemical reactions in human cells
_____ **5.** Miller and Urey	
_____ **6.** Cech and Altman	**b.** proved that organic molecules can be assembled spontaneously under laboratory conditions that simulate those thought to exist on early Earth
	c. thought that photons from the sun and electrical energy from lightning produced chemical reactions in molecules of nitrogen gas, water vapor, methane, hydrogen gas, and ammonia to form complex organic molecules

Origin of the First Cells

Mark each statement below *T* if it is true and *F* if it is false.

_____ **7.** Scientists think that tiny spheres of lipid were the first stage in the development of the cell.

_____ **8.** Life began when lipid spheres were able to transfer to offspring the ability to survive longer.

_____ **9.** Most scientists believe that after simple cells existed, RNA evolved to ensure the safety of the hereditary information.

_____ **10.** Scientists have recently been able to create life from nonliving substances in the laboratory.

■ Section 11-2: *Early Life in the Sea* page 195

Earliest Life: Bacteria

Write the term that is being described in each space provided.

1. The most common bacteria today are _____ .

2. Bacteria believed to be the direct ancestors of eukaryotes are _____ .

3. Photosynthetic bacteria responsible for the oxygen content of today's atmosphere

are _____ .

Dawn of the Eukaryotes

Match the choices from column B with the kingdoms in column A, then write your answers in the space provided. (Each kingdom will have two answers.)

Column A	Column B
_____ **4.** eubacteria	**a.** oldest living things
_____ **5.** archaebacteria	**b.** most diverse
_____ **6.** protists	**c.** multicellular
_____ **7.** plants	**d.** prokaryotic
_____ **8.** animals	**e.** eukaryotic
_____ **9.** fungi	

Life Blooms in the Ancient Seas

Read each question, and write your answer in the space provided.

10. Most major groups of multicellular organisms that survive today originated in what

time period? _____

11. What are the major groups of each kingdom called?_____

12. Describe what has been found in the rock formation called the Burgess Shale in

Canada. _____ _____

13. How many mass extinctions have occurred during the history of life on Earth?

_____

■ Section 11-3: *Invasions of the Land* page 198

The Importance of Ozone

Place a check mark next to each accurate statement in the space provided.

_____ **1.** Prior to 400 million years ago, life was limited to the oceans.

_____ **2.** The chemical formula for ozone is N2.

_____ **3.** Ozone blocks out ultraviolet rays, making it safe to live on Earth.

Plants and Fungi Colonize Land

Complete each statement by writing the correct word in each space provided.

4. The mutualistic association between the roots of plants and fungi is called

_____.

5. _____ is a relationship in which both partners benefit.

Invasion of the Arthropods

Read the question, and write your answer in the space provided.

6. What were the first animals to leave the water and live on land?

_____

_____

The Drifting Continents

Mark each statement below *T* if it is true and *F* if it is false.

_____ **7.** Figure 11-11 illustrates continental drift.

_____ **8.** Continental drift has been taking place for only the past 2 million years.

_____ **9.** The continents move about because of heat and pressure from deep in the Earth.

Insects Were the First Flying Animals

Complete each statement by underlining the correct word in the brackets.

10. Today, insects are the [smallest / largest] group of animals.

11. The relationship between insects and flowering plants is an example of [mutualism / mycorrhizae].

■Section 11-4 *Parade of Vertebrates*

Animals With Backbones

Choose the phrase from column B that best describes the term from column A, then write your answer in the space provided. (There is only one correct answer for each term.)

Column A	Column B
_____ **1.** chordate	**a.** jawless fish
_____ **2.** notochord	**b.** bony skeletons, jaws, and gills
_____ **3.** vertebrates	**c.** group whose members have a notochord
_____ **4.** lamprey	**d.** examples include snakes, lizards, and turtles
_____ **5.** sharks	**e.** flexible rod that extends along the back
_____ **6.** bony fishes	**f.** regulate body heat by absorbing heat from the environment
_____ **7.** amphibians	**g.** have hair and produce milk to feed young
_____ **8.** reptiles	**h.** chordates with a vertebral column
_____ **9.** mammals	**i.** cartilage skeletons
_____ **10.** endotherms	**j.** land vertebrates that must reproduce in moist places
_____ **11.** ectotherms	**k.** regulate body temperature through internal mechanisms

12 Human Evolution

Directed Reading

▌Section 12-1: *Primates*

Evolution of Primates

Place a check mark next to each accurate statement in the space provided.

_____ **1.** All primates have six flexible fingers.

_____ **2.** All primates have hair and nurse their offspring with milk.

_____ **3.** All primates live in trees.

_____ **4.** All primates have opposable thumbs.

_____ **5.** All primates possess binocular vision.

_____ **6.** All primates see well at night because of disproportionately large eyes.

Anthropoids Are Day-Active Primates

Complete each statement by writing the correct word in each space provided.

7. We know that primates became day-active because they had smaller

_____ _____ than did the earlier prosimians.

8. It is believed that _____ developed color vision around
35 million years ago.

9. The _____ of anthropoids are larger than those of prosimians.

10. Monkeys such as those shown in Figure 12-3 take care of their young longer than

most other mammals, except for _____ and _____ .

Apes

Read each question, and write your answer in the space provided.

11. What kind of ape is most distantly related to humans? _____

12. What kind of ape is most closely related to humans? _____

13. How is the answer to question 12 supported by scientific studies of DNA and

nucleotide sequences? _____

_____

■ Section 12-2 *Evolutionary Origins of Humans* page 218

Searching for the Fossil Relatives of Humans

Mark each statement below *T* if it is true and *F* if it is false.

_____ **1.** Humans and chimpanzees have a 98.4 percent genetic similarity but differ substantially in physical appearance.

_____ **2.** Early hominid fossils are plentiful and easy to find.

_____ **3.** Only hominids who died near swamps, lakes, or mud pits could become fossils.

_____ **4.** Hominid fossils are typically found in shallow, sandy soil.

Characteristics of the Earliest Hominids

5. Darwin predicted that hominid fossils would be found in _____ because chimpanzees and gorillas lived there.

6. By viewing the skeletons in Figure 12-2, it is evident that Australopithecines were

_____ when they walked.

7. The Australopithecine's brain was _____ than that of the chim-

panzee and _____ than that of the modern human.

Branches of the Hominid Evolutionary Tree

Complete the chart below.

	Hominid name	Characteristics	Year discovered	Location discovered
8.	A. africanus			
9.	A. robustus			
10.	A. boisei			
11.	A. afarensis			

■ Section 12-3: *The First Humans* page 223

Evolution of the Genus *Homo*

Read each question, and write your answer in the space provided.

1. Name the three members of genus *Homo*. _____

_____

_____

_____

2. Use Figure 12-13 to contrast the two hypotheses regarding the ancestors of humans.

_____

_____

_____

Our African Origins

Complete each statement by writing the correct word in each space provided.

3. Two types of remains that have been unearthed in the search for human origins

are _____ and _____.

4. The brain size of *Homo habilis* is about _____ times as large as that
of *A. afarensis*.

5. The major characteristics of *Homo habilis* are _____ brains,

_____ stature, and the use of _____.

6. The brain of *Homo erectus* was _____ cm^3 larger than that
of *Homo habilis*.

7. *Homo erectus* originated in _____.

8. By at least 1 million years ago, *Homo erectus* had migrated into _____

and _____.

The Origin of *Homo Sapiens*

Read each question, and write your answer in the space provided.

9. What does mitochondrial DNA analysis show about the evolution of the species *Homo sapiens*?

_____

_____

_____

10. Describe the lifestyle of Neanderthals.

_____

_____

_____

11. How long ago did *Homo sapiens,* as identical to modern humans, appear?

_____

_____

_____

12. What characteristic has marked the evolution of modern humans?

_____

_____

_____

13. How have humans undergone a cultural evolution?

_____

_____

_____

13 Animal Behavior

CHAPTER

Directed Reading

■ **Section 13-1:** *Evolution of Behavior*   page 233

What Is Behavior, and How Is It Studied?

Complete each statement by writing the correct word in each space provided.

1. An action or series of actions performed by an animal in response to a stimulus is

   called _____.

2. When scientists studying an animal behavior ask "how" questions, they are concerned

   with the _____ of the behavior.

3. When scientists studying an animal behavior ask "why" questions, they are concerned

   with the _____ for the behavior.

How Natural Selection Shapes Behavior

Mark each statement below *T* if it is true and *F* if it is false.

_____ 4. When a male lion takes over a pride, it kills all of the cubs.

_____ 5. Female lions will breed even when they are still caring for their cubs.

_____ 6. If a male lion kills all of the young cubs in a pride, the female lions will mate
            with him and have his offspring.

_____ 7. Natural selection favors traits that benefit groups or species.

The Genetic Basis of Behavior

Read each question, and write your answer in the space provided.

8. What does Figure 13-3 indicate about the nest-building behavior of the lovebird

   species shown? _____

   _____

   _____

9. How did the hybrids of the two lovebird species carry their nesting material?

   _____

   _____

10. Look at Figure 13-4. What was the result of William Cade's experiment on the chirping time of male crickets? _____

_____

_____

Learned Behaviors Are Modified by Experience

Choose the scientist from column A who conducted the experiments described in column B, and write your answer in the space provided.

Column A	Column B
a. Konrad Lorenz	_____ **11.** taught dogs to associate the ringing of a bell with food
b. Ivan Pavlov	_____ **12.** studied trial-and-error learning in rats
c. B. F. Skinner	_____ **13.** discovered that imprinting in young ducks and geese takes place during a short period following hatching

Genes and Learning Interact to Shape Behaviors

Complete each statement by underlining the correct word in the brackets.

14. When Robert Tryon bred the maze-bright rats with one another, their offspring made [more / fewer] maze-running mistakes.

15. The graphs in Figure 13-8 indicate that the rats' ability to learn a maze is at least partly [hereditary / spontaneous].

16. When the two groups of rats were tested using a different maze, there was [a / no] difference in their maze-running ability.

▌Section 13-2 *Kinds of Behavior*
page 240

Categories of Behavior

Read the question, and write your answer in the space provided.

1. What are at least three types of animal behavior? Explain the function of each in the space provided. _____

_____

_____

_____

How Animals Communicate

Place a check mark next to each accurate statement in the space provided.

_____ **2.** Some fishes communicate through a sensitivity to weak electrical fields.

_____ **3.** Tungara frogs most effectively attract mates by using visual signals such as movement.

_____ **4.** Primates can communicate the identity of specific predators by using chemical signals.

_____ **5.** The ability of a child to rapidly learn a vocabulary of thousands of words seems to be genetically programmed.

Choosing a Mate

Complete each statement by writing the correct word in each space provided.

6. The _____ produced by a female silk moth attracts only males of her own species.

7. Figure 13-12 shows male animal traits that are favored by _____

_____.

8. The males of some species of worms, butterflies, and snakes gain an advantage in reproduction over other males of their species by sealing the females'

_____ _____ after mating.

The Evolution of Self-Sacrificing Behavior

Read each question, and write your answer in the space provided.

9. Write a sentence using the word *altruism* correctly.

_____

_____

_____

10. List at least two examples of altruistic behavior in animals.

_____

_____

_____

11. Using Figure 13-15 as a guide, briefly explain kin selection.

_____

_____

_____

_____

_____

<table>
<tr><td>CHAPTER</td></tr>
</table>

14 Ecosystems

Directed Reading

■ Section 14-1: *What Is an Ecosystem?*

page 253

State of Our World

Mark each statement below *T* if it is true and *F* if it is false.

_____ **1.** In the next year, humans will cut down or burn about 17 million hectares of forest.

_____ **2.** The population of Mexico will double in the next year.

Ecology and Ecosystems

Choose the phrase from column B that best describes the term in column A, and write your answer in the space provided.

Column A	Column B
_____ **3.** ecology	**a.** the physical location of a community
_____ **4.** community	**b.** the study of the interactions of organisms with one another and with their environment
_____ **5.** habitat	
_____ **6.** ecosystem	**c.** the measure of the number of species living in an ecosystem
_____ **7.** diversity	**d.** the organisms that live in a particular place
	e. a self-sustaining collection of organisms and their physical environment

Why Study Ecology?

Complete each statement about the study of ecology by underlining the correct word or phrase in the brackets.

8. It is very difficult to study an ecosystem because it can contain [thousands of interacting species \ eight or more trophic levels].

9. A model of an ecosystem [can usually / cannot] consider all of the factors that affect it.

10. Using a model of an ecosystem, ecologists can predict what will happen if some part of the real ecosystem is changed [with great accuracy / only as accurately as the information used to build the model].

Energy in Ecosystems

Complete the following chart.

	Organism	Producer or consumer	Herbivore, carnivore or omnivore	Trophic level
11.	Cows			
12.	Humans			
13.	Grass			

How Many Trophic Levels Can an Ecosystem Contain?

Mark each statement below *T* if it is true and *F* if it is false.

_____ **14.** When a mouse eats a plant, only 30 percent of the energy present in the plant's molecules goes into the mouse's molecules.

_____ **15.** When a snake eats a mouse, only 10 percent of the energy present in the mouse's molecules goes into the snake's molecules.

_____ **16.** An ecological pyramid shows the amount of energy found in each trophic level of an ecosystem.

_____ **17.** In the ecosystem represented in Figure 14-7, hawks are more numerous than mice.

■Section 14-2 *Cycles Within Ecosystems*   page 259

Nutrient Cycles

Read each question, and write your answer in the space provided.

1. What was the ecological question studied by Bormann and Likens?

_____

_____

_____

2. What did Bormann and Likens conclude from their studies of the effect of the cutting of trees and vegetation on the forest ecosystem?

_____

_____

_____

3. Can most organisms use the nitrogen present in the air to produce proteins? Use Figure 14-9 to explain your answer. _____

_____

_____

4. By what process does nitrogen in the air change to ammonia?

_____

_____

_____

Water Cycle

Mark each statement below *T* if it is true and *F* if it is false.

_____ **5.** Figure 14-10 shows that moisture returns to the atmosphere through evaporation.

_____ **6.** About 10 percent of the moisture that enters a tropical rain forest ecosystem passes through its plants and evaporates from their leaves.

_____ **7.** When forests are cut, the water cycle and nutrient cycles are broken.

Carbon Cycle

Complete each statement by writing the correct word in each space provided.

8. Figure 14-11 shows that cellular respiration by plants, animals, and decomposers can return carbon to the _____.

9. Humans are increasing the concentration of _____

_____ in the atmosphere by burning large amounts of fossil fuels.

10. When gases such as carbon dioxide retain the sun's heat, the atmosphere is warmed. This phenomenon is called the _____ _____.

■ Section 14-3 *Kinds of Ecosystems* page 263

Freshwater Ecosystems

Complete each statement by underlining the correct word in the brackets.

1. Freshwater ecosystems support a rich array of life, including [bacteria / plankton], the photosynthetic organisms at the base of aquatic food webs.

2. Freshwater ecosystems usually have three zones: the [deep / shallow] edge zone, the open-water [surface / underwater] zone, and the deep-water zone, where no [nutrient / light] enters.

Terrestrial Ecosystems

Choose the letter of the biome in column B that best matches the description in column A, and write your answer in the space provided.

<table>
<tr><td colspan="2" align="center">**Column A**</td><td align="center">**Column B**</td></tr>
<tr><td>_____</td><td>**3.** yearly precipitation of 75 to 250 cm</td><td>**a.** tropical rain forests</td></tr>
<tr><td>_____</td><td>**4.** rich soil with tall, dense grasses</td><td>**b.** savannas</td></tr>
<tr><td>_____</td><td>**5.** more species than any other biome</td><td>**c.** desert</td></tr>
<tr><td>_____</td><td>**6.** inhabitants must conserve water to survive</td><td>**d.** temperate grasslands</td></tr>
<tr><td>_____</td><td>**7.** ground is always frozen</td><td>**e.** deciduous forests</td></tr>
<tr><td>_____</td><td>**8.** cold, wet climate provides good conditions for growth of conifers</td><td>**f.** coniferous forests</td></tr>
<tr><td>_____</td><td>**9.** yearly precipitation of 90 to 150 cm</td><td>**g.** tundra</td></tr>
</table>

Ocean Ecosystems

Write the term that is being described in the space provided.

10. This ecosystem contains many fishes, and plankton is the primary producer.

_____

11. This small area contains the most diverse life in the ocean ecosystem.

_____

12. Photosynthesis cannot occur in this part of the ocean, but some bacteria have evolved

a way to make food without light. _____

<table>
<tr><td>CHAPTER</td></tr>
</table>

15 How Ecosystems Change | Directed Reading

■ Section 15-1 *Interactions Within Ecosystems*   page 273

Evolution and Ecosystems

Read each question, and write your answer in the space provided.

1. What two major factors influence the evolution of a species? _____

_____

_____

2. What is coevolution? Give an example of this process. _____

_____

_____

Coevolution Shapes Species Interactions

Complete each statement by writing the correct word in each space provided.

3. The evolution of flowering plants and the insects that transport their male gametes is

an example of _____.

4. Animals that carry pollen from flower to flower are called _____.

5. To attract the flies that aid its pollination, the flower in Figure 15-2 releases an odor

that smells like _____ _____.

Avoiding Being Eaten: Plants and Herbivores

Mark each statement below *T* if it is true and *F* if it is false.

_____ **6.** Some plants use physical defenses such as thorns and tough leaves
for protection against herbivores.

_____ **7.** Very few plant species produce chemicals that protect them against herbivores.

_____ **8.** Every group of closely related plant species has a unique battery of physical
defenses, which are the most crucial of all plant defenses.

_____ **9.** Some herbivores have evolved ways to overcome the chemical defenses
of plants.

Three Types of Close Species Interactions

Write the term that is being described in each space provided.

10. In this symbiotic relationship, all participating species benefit.

_____

11. In this relationship, one species benefits and the other is not visibly affected.

_____

12. This is an antagonistic relationship between a parasite and its host.

_____

13. An example of this relationship is the lichen, a partnership between a fungus

and a green alga. _____

14. The crusty barnacles on the back of a gray whale illustrate this type of relationship.

_____

15. The single-celled organism that causes malaria when living in the blood of a human is

an example of this type of relationship. _____

■ Section 15-2 *Ecosystem Development and Change*   page 278

Ecosystem Lifestyles

Mark each statement below *T* if it is true and *F* if it is false.

_____ **1.** An organism's niche is the sum of its interactions with only its physical environment.

_____ **2.** The niche of an organism includes the climate it prefers, the time of day it feeds, the food it likes to eat, and the time of year it reproduces.

_____ **3.** The total niche that an organism could potentially use within an ecosystem is that organism's fundamental niche.

Competing Organisms Coevolve

Complete each statement by underlining the correct word in the brackets.

4. The two species of barnacles in Figure 15-9 were [cooperating / competing] with each other for living space on the rocky coast.

5. The part of a fundamental niche that an organism actually occupies is called

its [realized / assumed] niche.

6. Competitive exclusion results in the [restriction / death] of one of the two

species in competition for the same resource.

7. Competitive exclusion is rare in most ecosystems because natural selection

tends to favor evolutionary changes that [increase / decrease] competition.

Competition and Ecosystem Development

Place the following stages of a developing ecosystem in correct sequential order by renumbering the steps in the spaces provided.

_____ **8.** Small, fast-growing plants that are specialized for life under harsh conditions move into the habitat.

_____ **9.** After a long time, a community that is resistant to change develops.

_____ **10.** The eruption of a volcano forms a new, bare rock surface. A nearly empty habitat is established.

_____ **11.** Once the ground becomes more hospitable, other plants out-compete and replace the original inhabitants.

Ecosystem Stability

Mark each statement below *T* if it is true and *F* if it is false.

_____ **12.** More-diverse ecosystems are less stable than less-diverse ecosystems.

_____ **13.** A less-diverse ecosystem contains a less-complex web of interactions between species than does a more-diverse ecosystem.

_____ **14.** A keystone species has a niche that affects many other species in the ecosystem and cannot be easily replaced.

_____ **15.** The loss of a keystone species has very little effect in a diverse ecosystem.

Why Are Some Ecosystems More Diverse Than Others?

Mark each statement below *T* if it is true and *F* if it is false.

16. The following conditions lead to a more-diverse ecosystem:

_____ **a.** The ecosystem occupies a large area.

_____ **b.** The ecosystem is subdivided into isolated communities.

_____ **c.** The ecosystem contains a wide variety of physical habitats.

_____ **d.** The ecosystem has a very short growing season.

_____ **e.** The ecosystem has plenty of sunlight, rainfall, and warm temperatures.

_____ **f.** The ecosystem has an unstable climate.

How Humans Disrupt Ecosystems

Read the question, and write your answer in the space provided.

17. What are the three principal ways that humans disrupt ecosystems? Give an example

of each. _____

_____

_____

_____

_____

_____

_____

_____

_____

_____

16 The Fragile Earth

CHAPTER

Directed Reading

Section 16-1: *Planet Under Stress*

Humans Have Damaged the Environment

Mark each statement below *T* if it is true and *F* if it is false.

_____ **1.** Damage done to any one ecosystem can have ill effects on other ecosystems.

_____ **2.** Americans throw away almost 10 pounds of garbage a day.

_____ **3.** Pollution is anything potentially harmful that humans add to the environment.

_____ **4.** A carcinogen is a substance that can cause malaria.

Pollution's Toll

Complete each statement by writing the correct word in each space provided.

5. _____ percent of Poland's deep wells are polluted.

6. In 1989, the oil tanker _____ _____ ran aground off the coast of Alaska and spilled oil that polluted about 1,600 km of coastline.

7. When the sulfur released by smokestacks combines with water vapor in the upper atmosphere, the water vapor later condenses and forms _____ _____ .

8. An acidic solution has a higher concentration of _____ ions than does pure water.

Destroying the Ozone Layer

On the line at the left, write the letter of the answer that best completes each sentence.

_____ **9.** The series of photographs in Figure 16-4 shows
　　a. the hole in the ozone layer over the South Pole.
　　b. the hole in the ozone layer over North America.
　　c. the hole in the ozone layer over the North Pole.
　　d. how the ozone layer is regenerating itself.

_____ **10.** The primary cause of the destruction of ozone in the upper atmosphere is
　　a. a higher concentration of sulfuric acid in acid precipitation.
　　b. higher carbon dioxide levels in the atmosphere.
　　c. chlorofluorocarbons reaching the upper atmosphere.
　　d. increased amounts of ultraviolet radiation reaching our planet.

_____ **11.** Exposure to higher levels of ultraviolet radiation can lead to
    a. more productive food crops.
    b. larger animal populations.
    c. warmer summers and cooler winters.
    d. skin cancer and cataracts.

Global Warming

Read each question and write your answer in the space provided.

12. How have large quantities of carbon dioxide been released into our atmosphere?

_____

_____

13. What causes the warming of Earth's atmosphere? _____

_____

_____

14. How does Figure 16-5b illustrate the cause of global warming? _____

_____

_____

15. By what degree range is the world's average annual temperature expected to increase

by the year 2100? _____

Section 16-2: *Meeting the Challenge*
page 294

Reducing Pollution

Place a check mark next to each accurate statement.

_____ **1.** Industries that release pollution are required to purchase pollution permits.

_____ **2.** Laws have been passed to restrict the amount of sulfur that can be released by power plants.

_____ **3.** Private industries are required by law to clean up any pollution released into the environment.

Finding Enough Energy

Mark each statement below *T* if it is true and *F* if it is false.

_____ **4.** Oil and natural gas are nonrenewable resources.

_____ **5.** Nuclear power plants that are now being used to provide us with energy pose no threats to our safety.

_____ **6.** Solar energy and wind power are both examples of renewable resources.

_____ **7.** Only about 10 percent of the electricity used in the United States and Canada is wasted through inefficient appliances.

Conserving Nonrenewable Resources

Complete each statement by writing the correct word in each space provided.

8. The intense cultivation of crops causes an enormous loss of _____ each year.

9. Humans are rapidly depleting the supply of water trapped beneath the soil known

as _____ _____ .

10. During the past 20 years, approximately _____ of the world's tropical rain forests have been destroyed.

11. As the tropical rain forests are destroyed, many species of living things will become

_____ .

12. The plant in Figure 16-9a is the source of two drugs used to treat

_____ .

The Deeper Problem: Population Control

On the line at the left, write the letter of the answer that best completes each sentence.

_____ **13.** The human population started to expand explosively around 1650 because
   a. fertility rates in humans increased dramatically.
   b. of a more plentiful supply of food.
   c. of the spread of better sanitation and improved medical care.
   d. deaths from predators decreased dramatically.

_____ **14.** The world's annual population growth rate is currently
   a. 1.5 percent.
   b. 5 percent.
   c. 15 percent.
   d. 25 percent.

_____ **15.** The world's population is expected to peak by the middle of the next century at somewhere between
   a. 8 million and 13 million people.
   b. 2 billion and 4 billion people.
   c. 5 billion and 10 billion people.
   d. 8 billion and 13 billion people.

■Section 16-3: *Solving Environmental Problems*   page 300

Environmental Problems Can Be Solved

Complete each statement by underlining the correct word or phrase in the brackets.

1. It is important to remember that environmental problems are [solvable/unsolvable].

2. [Few/Many] endangered species are better off today than they were in 1970.

3. Pollution controls [have been/have not been] particularly successful.

4. Private firms and public agencies are estimated to be spending [twice/five times] as much on pollution control as was being spent in 1970.

Steps Toward Saving the Environment

Use the plan of action in Table 16-1 to develop a brief list of what you can do to help clean a pond that is polluted with runoff from a nearby industry.

5. Assessment _____

_____

_____

_____

6. Risk-analysis _____

_____

_____

_____

7. Public education _____

_____

_____

_____

8. Political action _____

_____

_____

_____

9. Follow-through _____

_____

_____

_____

What You Can Contribute

Read each question, and write your answer in the space provided.

10. What percentage of the world's population lives in the United States? How much of the

world's energy does the United States use? _____

_____

_____

11. List four things you can do to reduce your consumption of energy. _____

_____

_____

_____

_____

12. List four things you can do to help reduce waste and pollution. _____

_____

_____

_____

_____

Why Learning About Ecology Is Important

On the line at the left, write the letter of the answer that best completes the sentence.

_____ **13.** The most important way you can contribute to solving environmental
problems is to
    a. refuse to use any nonrenewable resources.
    b. make a serious effort to understand the environment.
    c. move to a region where heating and cooling needs are minimal.
    d. refuse to support candidates for office who are not committed
environmentalists.

17 Classifying Living Things | Directed Reading

CHAPTER

▌Section 17-1: *The Need for Naming* page 313

The Importance of Scientific Names

Complete each statement by writing the correct word in each space provided.

1. Each kind of organism on Earth is assigned a unique two-word _____

_____ .

2. The scientific name for modern human beings is _____

_____ .

3. The use of scientific names allows scientists to exchange information about a(n)

_____ no matter what language they speak.

What's in a Scientific Name?

Mark each statement below *T* if it is true and *F* if it is false.

_____ **4.** The first word of a scientific name is the name of the genus to which the organism belongs.

_____ **5.** Each different kind of organism is called a species.

_____ **6.** Two organisms can have the same scientific name.

_____ **7.** Scientific names are written in French.

_____ **8.** The Swedish botanist Carl Linnaeus developed the modern system of naming organisms.

_____ **9.** The scientific name for human beings can be written as *H. Sapiens.*

(continued on the next page)

■ Section 17-2: *Classification: Organizing Life*   page 317

Classification of Living Things

Choose the statement from column B that best matches the term in column A, and write your answer in the space provided.

	Column A	Column B
_____	**1.** taxonomy	**a.** level that contains families that are alike
_____	**2.** family	**b.** science of classifying living things
_____	**3.** order	**c.** level that includes similar phyla
_____	**4.** class	**d.** classes are united in this level of classification
_____	**5.** phylum	**e.** level of biological classification made up of similar genera
_____	**6.** kingdom	**f.** similar orders are collected at this level

Classification and Evolution

Complete each statement by writing the correct word in each space provided.

7. The more similarities two organisms have, the more recently they shared

a _____ _____ .

8. The shark and the dolphin in Figure 17-6 provide an example of

_____ _____ .

9. If two different organisms evolve similar structures independently, these structures are

called _____ _____ .

Methods of Taxonomy

Identify the following statements as descriptions of caldistics (C) or phenetics (P), the two methods of classifying organisms.

_____ **10.** Organisms are assigned to a group because they share derived characteristics not found in other organisms.

_____ **11.** Organisms are classified according to their degree of similarity.

_____ **12.** Taxonomists try to determine the order in which evolutionary lines branched and to discover relationships among organisms.

_____ **13.** The relationships among organisms are not reconstructed.

Taxonomy and Technology

Read each question, and write your answer in the space provided.

14. What technological advances have allowed biologists to have a better look at how

organisms are related? _____

_____

_____

_____

15. Explain the following statement: DNA acts as a "molecular clock."_____

_____

_____

_____

16. What can a comparison of DNA sequences in two species reveal? _____

_____

_____

_____

What Is a Species?

Place a check mark next to each accurate statement.

_____ **17.** A species is a group of organisms able to interbreed with each other to produce fertile offspring that usually do not reproduce with members of any other groups.

_____ **18.** A species is a level in the biological classification system containing very similar organisms.

_____ **19.** Because they can mate, horses and zebras belong to the same species.

_____ **20.** The term *species* comes from the Latin word that means "kind."

_____ **21.** A group of individuals that share at least four inherited characteristics that are not found in other similar organisms are a species.

Section 17-3: *Six Kingdoms*

Six-Kingdom System

Complete each statement by underlining the correct word or phrase in the brackets.

1. Linnaeus classified everything into one of two kingdoms, kingdom Animalia or kingdom [Plantae/Fungi].

2. Most modern biologists use a [five/six]-kingdom classification system.

3. The organism in Figure 17-10 is classified as a(n) [plant/animal].

The Six Kingdoms of Living Things

Mark each statement below *T* if it is true and *F* if it is false.

_____ **4.** Archaebacteria and eubacteria evolved from a common ancestor about 4 billion years ago.

_____ **5.** All multicellular eukaryotes not classified as plants, animals, or fungi are assigned to the kingdom Protista.

_____ **6.** The organism shown in Figure 17-13 is a member of the kingdom Fungi.

_____ **7.** The kingdom Plantae contains only single-celled organisms that obtain their nutrients by photosynthesis.

_____ **8.** Plant cells have cell walls made of cellulose.

_____ **9.** The first members of the kingdom Animalia evolved in the ocean.

_____ **10.** Specialized cells in animals can photosynthesize.

18 Bacteria and Viruses

CHAPTER

Directed Reading

▌**Section 18-1** *Bacteria*   page 331

Bacteria Are Small and Successful

Complete each statement by underlining the correct word or phrase in the brackets.

1. Bacteria live [almost everywhere/only in hostile habitats] on our planet.

2. Figure 18-1 shows the [five/three] general shapes of bacteria.

3. All bacteria are [prokaryotes/eukaryotes].

4. A [Gram-positive/Gram-negative] bacterium's cell wall retains the purple color of gram staining.

5. Gram-negative bacteria are [killed/unaffected] by many antibiotics.

6. Bacteria reproduce [by splitting in two/through mitosis].

How Bacteria Obtain Nutrition

Read each question and write your answer in the space provided.

7. What is a major reason for the success of bacteria? _____

_____

_____

8. Where do autotrophic bacteria such as those found in Figure 18-3 obtain their energy?

_____

_____

9. In the absence of sunlight, what source of food can chemosynthetic bacteria utilize?

_____

_____

10. From what sources do heterotrophic bacteria obtain their food?

_____

_____

▍Section 18-2: *How Bacteria Affect Humans*   page 336

Beneficial Bacteria

Complete each statement by writing the correct word in each space provided.

1. Some bacteria serve as _____, organisms that return nutrients to the environment by breaking down organic matter.

2. Nitrogen-fixing bacteria transform atmospheric nitrogen into _____, a nitrogen compound that plants can use.

3. Most insulin needed by diabetics in the United States today is produced

by _____ .

Bacteria and Disease

Mark each statement below *T* if it is true and *F* if it is false.

_____ **4.** Pathogenic bacteria are harmful because they can damage their host's tissues.

_____ **5.** According to the information in Table 18-1, typhus is spread by fleas.

_____ **6.** Bacteria cause tooth decay.

_____ **7.** Salmonella bacteria are typically spread through fresh and frozen foods.

_____ **8.** Bacterial contamination of food can be avoided through the appropriate use of heat and cold.

_____ **9.** Three million people die annually from the airborne bacterial disease tuberculosis.

_____ **10.** Lyme disease is caused when a tick releases bacteria into the digestive system of a human.

Controlling Bacterial Diseases

On the line at the left, write the letter of the answer that best completes each sentence.

_____ **11.** Cholera bacteria are primarily spread through
    a. contaminated hypodermic needles.
    b. mosquitoes and fleas.
    c. contaminated drinking water.
    d. coughing and sneezing.

_____ **12.** Figure 18-11 shows how
    a. a vaccine for whooping cough is manufactured and used.
    b. whooping cough kills 25 million people per year.
    c. contaminated drinking water may be purified.
    d. children dislike injections of any type.

_____ **13.** In 1928, Alexander Fleming accidentally discovered the antibiotic
    a. tetracycline.
    b. streptomycin.
    c. AZT.
    d. penicillin.

_____ **14.** Which of the following statements is **not** true about antibiotic-resistant bacteria?
    a. Failure to complete a full course of antibiotics may create them.
    b. Scientists have won the battle against all antibiotic-resistant bacteria.
    c. Their resistance may be passed along to offspring.
    d. They may not be affected by certain antibiotics.

▌Section 18-3: *Viruses*

page 345

What Is a Virus?

Mark each statement below *T* if it is true and *F* if it is false.

_____ **1.** A typical virus is composed of a core of genetic material surrounded by a protein coat.

_____ **2.** All viruses have only DNA as their genetic material.

_____ **3.** The small strands of viral genetic material contain only a few genes.

_____ **4.** Most biologists consider viruses as living cells.

How Viruses Reproduce

Place the following events in the reproduction of a virus in the correct order by numbering the steps in the blank provided.

_____ **5.** Viral genetic material is incorporated into the cell's chromosomes.

_____ **6.** The virus enters the host cell.

_____ **7.** Newly formed viruses break out of the cell and continue the cycle of infection.

_____ **8.** Viral cell marker proteins bind to a specific receptor protein on the cell's surface.

_____ **9.** The host cell begins to form viral proteins and genetic material to make new viruses.

Diseases Caused by Viruses

Complete each statement by underlining the correct word or phrase in the brackets.

10. Antibiotics [can/cannot] be used to treat viral diseases.

11. Vaccines against viral diseases contain [bacteria/viruses] that have been made harmless.

12. Smallpox was [eradicated/widespread] in the United States by the year 1949.

13. Filoviruses such as the Ebola virus are believed to be spread by [chimpanzees/ unidentified animals].

14. Vaccines have not been effective for HIV, cold viruses, and flu viruses because their cell surface proteins [mutate frequently/remain the same over time].

15. Table 18-2 provides examples of viral diseases, including such fatal diseases as [yellow fever and the Ebola virus/chickenpox and mumps].

19 Protists

CHAPTER 19

Directed Reading

▮Section 19-1 *What Is a Protist?*

page 357

Characteristics of Protists

Mark each statement below *T* if it is true and *F* if it is false.

_____ **1.** All protists are single-celled prokaryotes.

_____ **2.** Protist cells have a nucleus that contains chromosomes.

_____ **3.** Protist cells contain membrane-bound organelles such as chloroplasts and mitochondria.

_____ **4.** All protists are heterotrophic.

_____ **5.** The amoeba in Figure 19-4 is a protist that uses cilia and flagella to move.

_____ **6.** Most protists can sense light, touch, and chemicals.

_____ **7.** Protists reproduce both sexually and asexually.

_____ **8.** The protist in Figure 19-5 only reproduces sexually.

Evolutionary Relationships Among Protists

Complete each statement by underlining the correct word or phrase in the brackets.

9. Because it contains both mitochondria and chloroplasts, Euglena is a union of [two/three] evolutionary lines.

10. Protists are the ancestors of the major [multicellular eukaryotic kingdoms/ single-celled prokaryotic kingdoms].

▮Section 19-2: *Protist Diversity*

page 360

Classification of the Protists

Mark each statement below *T* if it is true and *F* if it is false.

_____ **1.** The protists in Figure 19-6 are unicellular heterotrophs.

_____ **2.** There is a greater variety of sizes and structures in the kingdom Protista than in any other kingdom.

_____ **3.** Among the eukaryotes, the kingdom Protista is the only one that includes both heterotrophs and autotrophs.

Autotrophic Protists

Choose the term from column B that best matches the statement in column A, and write your answer in the space provided.

Column A	Column B
_____ **4.** occur as kelp forests off the coast of California	**a.** phytoplankton
_____ **5.** considered to be the ancestors of plants	**b.** diatoms
_____ **6.** have glasslike shells sometimes used to make detergents and fertilizers	**c.** dinoflagellates
_____ **7.** cause the red tides that are poisonous to fish	**d.** green algae
_____ **8.** found near the surface because they require light	**e.** brown algae
_____ **9.** distinctive pigments allow photosynthesis in deep waters	**f.** red algae

Heterotrophic Protists

Complete each statement by underlining the correct word or phrase in the brackets.

10. Most [autotrophic/heterotrophic] protists are either carnivores or herbivores.

11. The cell walls, composition, and structure of slime molds and water molds [differ from/are the same as] those of fungi.

12. Protists known as zoomastigotes are probably the ancestors of [animals/plants].

13. Some scientists believe that recent molecular analysis of DNA indicates that ciliates belong in [a separate kingdom of their own/the kingdom Protista].

▪ Section 19-3: *Diseases Caused by Protists*   page 366

Protists and Disease

Complete each statement by writing the correct word in each space provided.

1. Disease-causing protists are transmitted primarily by _____ and

_____ .

2. Malaria kills more than _____ _____ people each year.

3. Figure 19-13 shows the life cycle of the protist that causes malaria, a parasite called

_____ .

4. The bitter chemical _____ and its derivatives have been used since the 1600s to treat and prevent malaria.

5. Malaria was eradicated from the southern United States by eliminating the

mosquitoes' breeding grounds and the use of powerful _____ .

6. Human efforts to eliminate malaria have been thwarted by _____
advances by both the Plasmodium parasite and the Anopheles mosquito.

7. The Plasmodium parasite has developed _____ to some of the drugs that previously controlled it.

8. Diseases caused by protists are most likely to affect people who live in countries that

are less _____ .

9. The disease in Table 19-1 for which both cats and humans can serve as hosts

is _____ .

10. Trypanosoma can cause both _____ _____ and

_____ _____ .

20 Fungi and Plants

CHAPTER

Directed Reading

Section 20-1 *Fungi*

Characteristics of Fungi

Complete each statement by writing the correct word in each space provided.

1. Because most fungi are immobile and appear to be rooted, biologists once placed

   them in the _____ kingdom.

2. Fungi do not contain _____ and therefore cannot capture energy
   from sunlight.

3. The body of a fungus is made up of long, thin filaments called

   _____ .

4. The cells of fungi have cell walls made of a polysaccharide called

   _____ .

5. Fungi _____ their food outside their bodies, then _____
   the nutrients through the hyphae.

6. When a fungal _____ lands on a food source with enough moisture,
   it rapidly grows into a hypha.

Kinds of Fungi

On the line at the left, write the letter of the term that best matches the description.

_____ 7. No mode of sexual reproduction has been observed for
   this division of fungi.

_____ 8. These fungi include yeasts and morels.

_____ 9. The common bread mold is a member of this division of fungi.

_____10. Saclike reproductive structures contain the spores in this
   division.

_____11. Mushrooms are produced by some members of this division.

_____12. One member of this division is the fungus that produces
   penicillin.

a. ascomycetes

b. zygomycetes

c. deuteromycetes

d. basidiomycetes

Fungi in Nature

Read each question, and write your answer in the space provided.

13. Explain the symbiotic relationship known as mycorrhizae. _____

_____

_____

14. What role do fungi play in the damage to forests by acid rain? _____

_____

_____

15. A lichen is a mutualistic association between a fungus and an alga. Describe how this

association works. _____

_____

_____

Fungi and Humans

Read each question, and write your answer in the space provided.

16. What is the scientific name of the ascomycete commonly known as baker's yeast?

_____

_____

17. What is the scientific name for the fungus that led to the development of penicillin?

_____

_____

18. The success rate of an important medical procedure has been increased dramatically
by the use of cyclosporine. Name the medical procedure, and tell how cyclosporine

works. _____

_____

_____

■ Section 20-2: *Early Land Plants*

Characteristics of Plants

Complete each statement by underlining the correct word or phrase in the brackets.

1. Plants are [photosynthetic/chemosynthetic] and have cell walls made of [chitin/cellulose].

2. Figure 20-7 shows how plants overcame the challenges of surviving [on land/ underwater].

3. Plants with a system of internal tubes that are used to distribute water and nutrients are called [vascular/nonvascular] plants.

4. Most plants are covered with a waxy, waterproof layer called the [cell wall/cuticle] that reduces the evaporation of water from leaves and stems.

5. During a plant's life cycle, a haploid [gametophyte/embryo] alternates with a diploid [stalk /sporophyte].

Nonvascular Plants

Mark each statement below *T* if it is true and *F* if it is false.

_____ 6. The first plants on land were nonvascular plants that lived in moist, shady environments.

_____ 7. Nonvascular plants grow to be very tall, sometimes up to 80 m high.

_____ 8. The most widespread nonvascular plants are the mosses.

_____ 9. Fertilization in nonvascular plants can take place only when the gametophytes are dry.

_____ 10. Peat, made of the moss known as Sphagnum, is used as fuel in Ireland, Canada, and Siberia.

Evolution of Vascular Plants

Complete each statement by underlining the correct word or phrase in the brackets.

11. Vascular tissue enabled vascular plants to be [larger than/the same size as] nonvascular plants.

12. Ferns provide an example of [sporeless/seedless] vascular plants.

13. As seen in Figure 20-11, the dominant phase of a fern in the alternation of generations is the [sporophyte/gametophyte].

▌Section 20-3: *Seed Plants*

Vascular Plants With Seeds

Read each question, and write your answer in the space provided.

1. List two evolutionary changes that have enabled vascular plants to grow tall. _____

_____

_____

2. List three evolutionary changes that have allowed land plants to reproduce without water.

_____

_____

3. List the three main structures found in a seed. _____

_____

_____

4. List three seed structures that enable seeds to travel long distances and spread to new

areas. _____

_____

_____

Gymnosperms: Plants With Naked Seeds

Mark each statement below *T* if it is true and *F* if it is false.

_____ **5.** The word *gymnosperm* comes from the Greek words for "naked" and "seed."

_____ **6.** Gymnosperms produce their seeds inside fruits.

_____ **7.** Gymnosperms are able to reproduce without a film of water because of pollen.

_____ **8.** The spores produced by female cones develop into pollen grains.

_____ **9.** Pollen, which contains sperm, travels from where it is produced to the female
cones, where the eggs are located, during pollination.

Angiosperms: Flowering Plants

Complete each statement by writing the correct word in each space provided.

10. Angiosperms ensure the transfer of gametes through structures called

_____ .

11. Pollen is carried from one flowering plant to another by _____,

_____, or other _____ .

12. There are approximately _____ species of angiosperms.

13. As shown in Table 20-2, examples of the _____ family of angiosperms are broccoli, turnips, and cabbage.

21 Plant Form and Function | Directed Reading

▌Section 21-1: *The Plant Body*   page 391

Roots

Mark each statement *T* if it is true and *F* if it is false.

_____ **1.** Roots anchor a plant and absorb water and minerals from the soil.

_____ **2.** The roots of carrots and sweet potatoes have no special function.

_____ **3.** The root cap protects the tip of a root.

Shoots

Complete each statement by writing the correct word in each space provided.

4. The aboveground part of a plant is called the _____ .

5. Stems contain vascular tissue that transports substances between roots and

_____ .

6. Most cells in a leaf are full of _____, the sites of photosynthesis.

Plant Tissues

Choose the statement from Column B that best matches the term in Column A, and write the letter in the space provided.

Column A	Column B
_____ **7.** xylem	**a.** part of a plant that surrounds the vascular tissue
_____ **8.** phloem	**b.** causes plant bodies to thicken by producing new xylem and phloem
_____ **9.** ground tissue	
_____ **10.** epidermis	**c.** growth that occurs by the lengthening of a plant's roots and shoots
_____ **11.** primary growth	**d.** vascular tissue through which water and minerals flow
_____ **12.** secondary growth	
	e. layer of flattened cells that cover a plant's body
	f. vascular tissue through which sugars and other organic molecules flow

Monocots and Dicots

Place a check mark next to each accurate statement in the space provided.

_____ **13.** Grasses, orchids, wheat, and corn are monocots.

_____ **14.** The embryos of dicots have one cotyledon.

Section 21-2 *How Plants Function* page 396

How Nutrients Move Through Plants

Complete each statement by underlining the correct word or phrase in the brackets.

1. Water continually enters the roots of a plant through [osmosis/capillary action].

2. Water is pulled up the narrow tubes of a plant's xylem by [capillary action/gravity].

3. Over 90 percent of the water that enters a plant through its roots is lost through [translocation/transpiration].

4. Guard cells called [xylem/stomata] help plants control water loss.

5. Through [translocation/transpiration], organic molecules made during photosynthesis are moved to other parts of the plant.

Factors Regulating Plant Growth

Mark each statement below *T* if it is true and *F* if it is false.

_____ **6.** Plant hormones control the growth of plant shoots and roots.

_____ **7.** Plants grow toward the light because of a hormone called ethylene.

_____ **8.** Growth responses such as the bending of plants either toward or away from light, gravitational pull, or touch are called tropisms.

_____ **9.** The plants on the right of Figure 21-5 have been sprayed with water containing gibberellin.

_____ **10.** Ethylene is used commercially to keep fruit from ripening.

_____ **11.** The plants in Figure 21-6 are not sensitive to the change from daylight to darkness.

▌Section 21-3 *Reproduction in Flowering Plants* page 400

Architecture of a Flower

Complete each statement by underlining the correct word or phrase in the brackets.

1. The inner whorl of a flower where its male parts are located is called the [pistil/stamens].

2. The female parts of a flower are found in the [pistil/sepals].

Pollination and Fertilization

Choose the statement from Column B that best matches the term in Column A, and write the letter in the space provided.

Column A	Column B
_____ **3.** self-pollination	**a.** enlarged ovary of a flowering plant
_____ **4.** cross-pollination	**b.** process in which one sperm fuses with an egg to form a zygote and the other sperm fuses with two nuclei in the ovule to provide nutrition for the embryo
_____ **5.** double fertilization	
_____ **6.** fruit	
	c. transfer of pollen from one plant to another of the same species
	d. action that takes place when a flower pollinates itself

How Seeds Are Dispersed

Place a check mark next to each accurate statement in the space provided.

_____ **7.** Figure 21-9 shows the ways seeds can be dispersed.

_____ **8.** Seeds that pass through the digestive tracts of birds and mammals are damaged and, thus, cannot germinate.

_____ **9.** Water and oxygen must reach the embryo of a seed for germination to take place.

Plant Cell Growth and Differentiation

Study the following steps in the production of vegetables by tissue culture to determine the order in which they take place. Write the number of each step in the space provided.

_____ **10.** A new plant grows in the culture flask.

_____ **11.** Tissue is isolated from the tip of the stem.

_____ **12.** The plant is planted in the soil and grows to maturity.

_____ **13.** The stem is placed in a flask containing nutrients and hormones.

22 Plants in Our Lives

CHAPTER

Directed Reading

■ **Section 22-1** *Uses of Plants*

page 411

Important Grains

Read each question, and write your answer in the space provided.

1. What important grains are shown on the map in Figure 22-1? _____

_____

_____

2. Name at least three foods made from wheat.

_____

_____

3. Which of the grains in Table 22-1 contains the most protein? Which contain the most fiber?

_____

_____

4. What consumes 70 percent of the United States corn crop?

_____

_____

Other Food Plants

Choose the statement from Column B that best matches the term in Column A and write the letter in the space provided.

Column A	Column B
_____ **5.** essential amino acids	**a.** pressed cakes of cooked soybeans
_____ **6.** legumes	**b.** agricultural product that comes from a plant and is usually eaten with the main part of a meal
_____ **7.** tofu	**c.** members of the pea, or bean, family
_____ **8.** tubers	**d.** the amino acids not made by the human body
_____ **9.** vegetable	**e.** underground stems of potato plants

Uses of Wood

Mark each statement *T* if it is true and *F* if it is false.

_____ **10.** Wood is actually secondary xylem.

_____ **11.** Wood is the main fuel for about 10 percent of the people of the world.

_____ **12.** Paper comes from the cellulose fibers of such plants as cotton, bamboo, rice, sugar cane, and wheat.

Medicines From Plants

Place a check mark next to each accurate statement in the spaces provided.

_____ **13.** Aspirin, the world's most widely used drug, is made from willow leaves and bark.

_____ **14.** Digitalis is made of poppy fruits and regulates irregular heartbeat.

_____ **15.** Made from yew bark, taxol is used to reduce the size of cancerous tumors.

Other Plant Products

Answer the questions in the space provided.

16. What is the world's most important plant fiber? _____

_____

17. From what plant is turpentine made? _____

_____

18. What part of the rubber tree is used to produce natural rubber? _____

_____

(continued on the next page)

▌Section 22-2 *Growing Plants*

Soils and Plant Growth

Choose the statement from Column B that best matches the term in Column A, and write the letter in the space provided.

Column A	Column B
_____ **1.** macronutrients	**a.** ability of the soil to supply plants with nutrients
_____ **2.** micronutrients	**b.** crop that is grown for the sole purpose of plowing it back into the field
_____ **3.** loam	
_____ **4.** humus	**c.** nutrients, such as nitrogen and phosphorus, that plants need in relatively large amounts
_____ **5.** fertility	
_____ **6.** crop rotation	**d.** method of growing plants in nutrient solution
_____ **7.** green manure	**e.** soil that is a mixture of sand, silt, and clay
_____ **8.** composting	**f.** nutrients, such as zinc and iron, that plants need in very small amounts
_____ **9.** hydroponics	
	g. organic component of soil
	h. mixing organic debris and allowing it to begin the process of decomposition
	i. practice of alternating two or more crops in the same field from year to year

Choosing the Right Plants

Mark each statement below *T* if it is true and *F* if it is false.

_____ **10.** The map in Figure 22-18 divides the United States into plant hardiness zones.

_____ **11.** A soil test kit is used to test the pH of the soil.

Planting and Caring for Plants

Complete each statement by underlining the correct word or phrase in the brackets.

12. Seeds should be planted no deeper than they are [wide/long].

13. Trees and shrubs may be [pruned/mulched] to stimulate new growth and encourage flowering.

14. According to the information in Table 22-4, possible causes of holes in the leaves of plants are [caterpillars and pill bugs/fungi].

Plant Propagation

Read the question, and write your answer in the space provided.

15. What methods of vegetative propagation are described in Figure 22-21?

_____

_____

Plant Breeding

Study the following steps in the process of plant breeding to determine the order in which they take place. Write the number of each step in the space provided.

_____ **16.** screen offspring

_____ **17.** grow offspring

_____ **18.** select parent plants

_____ **19.** select desirable offspring

_____ **20.** cross parent plants

23 The Animal Body

CHAPTER

Directed Reading

Section 23-1 *The Origin of Tissues*   page 437

Animal Body Plans

Complete each statement by writing the correct word in each space provided.

1. The _____ _____ of an animal is its overall structure.

2. Figure 23-2 shows the animal body plans that are descended from

_____ ancestors.

Many Cells Instead of One: Sponges

Mark each statement *T* if it is true and *F* if it is false.

_____ **3.** The cells of a sponge are not specialized.

_____ **4.** The flagella of the choanocytes beat independently.

_____ **5.** **Exploration of a Sponge** shows that the body of a sponge has two layers of cells—an outer layer and an inner layer.

Tissues Enable Greater Cell Specialization: Cnidarians

Choose the statement from Column B that best matches the term in Column A, and write the letter in the space provided.

	Column A	Column B
_____	**6.** tissue	**a.** inner layer of cells formed by gastrulation
_____	**7.** gut	**b.** group of cells that are organized into a functional unit
_____	**8.** gastrulation	**c.** outer layer of cells formed by gastrulation
_____	**9.** endoderm	**d.** process of forming layers of cells
_____	**10.** ectoderm	**e.** internal passage through which food passes while being digested

Regularly Arranged Animals

Place a check mark next to each accurate statement in the space provided.

_____ **11.** Sponges are symmetrical.

_____ **12.** If an animal could be sliced in half to produce mirror-image body halves, the animal is referred to as symmetrical.

_____ **13.** Hydras, sea anemones, and jellyfish have radial symmetry.

Section 23-2 *Origin of Body Cavities* page 442

A Third Tissue Layer: Flatworms

Choose the statement from Column B that best matches the term in Column A, and write the letter in the space provided.

Column A	Column B
_____ **1.** bilateral symmetry	**a.** solid body construction of an animal in which the gut is completely surrounded by tissue and organs
_____ **2.** cephalization	**b.** "middle skin" of an animal that develops into muscle, reproductive organs, and circulatory vessels
_____ **3.** organs	**c.** animal's evolution of a definite head end
_____ **4.** organ system	**d.** collections of different kinds of tissue that are dedicated to one function
_____ **5.** mesoderm	**e.** type of body symmetry whereby an animal can be separated into nearly mirror-image halves when an imaginary line is drawn lengthwise down the middle of the body
_____ **6.** acoelomate	**f.** group of interrelated organs that carries out one or a few essential body functions

A One-Way Gut and a Body Cavity: Roundworms

Mark each statement below *T* if it is true and *F* if it is false.

_____ **7. Exploration of a Roundworm** shows that this animal has a one-way gut and an anus.

_____ **8.** The body cavity between the roundworm's gut and body wall is the pseudocoelom.

_____ **9.** Pseudocoelomate animals must either be very large or have bodies that provide long distances between organs and the body surface.

A Better Body Cavity: Mollusks

Complete each statement by writing the correct word in each space provided.

10. Most animals have a(n) _____, a fluid-filled body cavity that lies completely within the mesoderm.

11. One advantage of the coelem is that it allows for a longer and therefore more efficient

_____ tract.

12. Most coelomates have a(n) _____ _____ to overcome the barrier of the diffusion of nutrients presented by the solid tissue barrier surrounding the gut.

13. The primary features of the snail in **Exploration of a Mollusk** are its

_____ and its _____ _____ .

Section 23-3 *Four Innovations in Body Plan*   page 448

Segmented Worms: Annelids

Complete each statement by underlining the correct word or phrase in the brackets.

1. An example of [nonsegmentation/segmentation] in humans is the vertebral column.

2. The basic body plan of the animal in **Exploration of an Annelid** is a(n) [tube within a tube/exoskeleton of chitin].

Limbs and Skeletons: Arthropods

Read the questions, and write your answer in the space provided.

3. What are the key features of the animal in **Exploration of an Arthropod**?

_____

_____

4. Of what material is the exoskeleton of an arthropod made?

_____

_____

Two Patterns of Development

Place a check mark next to each accurate statement in the space provided.

_____ **5.** In protostomes, the first opening that forms in the embryo during gastrulation becomes the mouth.

_____ **6.** Echinoderms and chordates are protostomes.

_____ **7.** In Figure 23-15, tigers and hummingbirds are shown as examples of deuterostomes.

(continued on the next page)

Echinoderms

Complete each statement by writing the correct word in each space provided.

8. The key features of the animal shown in **Exploration of an Echinoderm** are

_____ _____ and _____ .

9. As an adult, the animal in **Exploration of an Echinoderm** has a

_____ radial symmetry.

The Most Successful Deuterostomes: Chordates

Mark each statement below *T* if it is true and *F* if it is false.

_____ 10. The lancelet shown in Figure 23-17 is a member of the same phylum
as humans: Chordate.

_____ 11. Eustachian tubes are the key features of the animal in **Exploration
of a Lancelet**.

_____ 12. The lancelet has no eyes, ears, or nose but instead has sensory cells
that cover the tentacles that surround the mouth.

24 Adaptation to Land

CHAPTER

Directed Reading

▌Section 24-1 *Leaving the Sea*

page 459

Which Animals Live on Land?

Mark each statement *T* if it is true and *F* if it is false.

_____ **1.** The only animals that have fully adapted to life on dry land are the arthropods and vertebrates.

_____ **2.** The thorny devil in Figure 24-1 has not adapted well to life on dry land.

Supporting the Body

Complete each statement by writing the correct word in each space provided.

3. The _____ provides support for arthropod and vertebrate bodies.

4. Figure 24-2 shows how changes in _____ _____ took place as reptiles evolved from fishes.

5. Compared with the limbs of amphibians, the limbs of the _____ in Figure 24-2d provide more support for body weight and enable the animal to run faster.

Hearing Airborne Sounds

Read the questions and write your answer in the space provided.

6. What do fishes use to sense physical disturbances in the water? _____

_____

_____

7. How do mammals detect sound? _____

_____

_____

8. What do crickets use to hear the sounds of other crickets? _____

_____

_____

▌Section 24-2 Staying Moist in a Dry World

Watertight Skin

Complete each statement by writing the correct word in each space provided.

1. In vertebrates, the watertight coating required to keep these animals from drying out

   is the _____ .

2. Made of chitin and protein, the _____ of terrestrial arthropods prevents water from leaving their bodies.

3. The _____ coating of a frog helps limit evaporation through their skin.

4. The development of _____ freed reptiles from living in a damp environment.

Gas Exchange

Place a check mark next to each accurate statement in the space provided.

_____ 5. Gills are used for gas exchange in animals that live on dry land.

_____ 6. Only fishes have gills.

_____ 7. The gills shown in Figure 24-9 provide a large surface area for gas exchange.

_____ 8. Gills would collapse if not supported by water.

Vertebrate Lungs

Complete each statement by underlining the correct word in the brackets.

9. Oxygen is brought into the lungs by [inhalation/exhalation] and diffuses through the lung lining into the [veins/capillaries].

10. The double-loop circulatory system illustrated in Figure 24-10b is found in all terrestrial [protists/vertebrates].

11. The double-loop circulatory system carries oxygenated blood through the body more [slowly/rapidly] than does the single-loop system.

How Terrestrial Arthropods Breathe

Mark each statement *T* if it is true and *F* if it is false.

_____ **12.** Insects breathe through tracheae.

_____ **13.** The tips of tiny branches in an insect's tracheae carry oxygen directly to its cells.

_____ **14.** The folds of a spider's book lungs provide only a small surface area for diffusion to take place.

Conserving Water While Getting Rid of Wastes

Choose the statement from Column B that best matches the term in Column A and write the letter in the space provided.

Column A	Column B
_____ **15.** ammonia	**a.** in vertebrates, the organ that acts as a filter of the blood
_____ **16.** urea	**b.** toxic waste excreted by freshwater animals
_____ **17.** uric acid	**c.** form of nitrogen waste excreted by most mammals
_____ **18.** kidneys	**d.** organs that filter blood in insects and spiders
_____ **19.** Malpighian tubules	**e.** solid, nontoxic nitrogen waste that can be excreted by animals with minimal water loss

∎Section 24-3 *Reproducing on Land*   page 472

Internal Fertilization

Choose the statement from Column B that best matches the term in Column A, and write the letter in the space provided.

Column A	Column B
_____ **1.** external fertilization	**a.** the type of egg shown in Figure 24-16
_____ **2.** internal fertilization	**b.** membrane that encloses the embryo of reptiles, birds, and mammals within a watery environment
_____ **3.** chorion	**c.** fertilization that takes place outside the body
_____ **4.** amnion	**d.** fertilization that takes place inside the female's body
_____ **5.** amniotic egg	**e.** watertight protective membrane that surrounds the embryos of reptiles
_____ **6.** yolk	**f.** source of nutrients for the developing reptile embryo

Eggs Without Shells

Read the questions, and write your answer in the space provided.

7. What are the disadvantages of the reptile or bird egg? _____

_____

_____

_____

8. Which mammals lay eggs? _____

_____

9. Where do mammal embryos develop? _____

_____

10. What activity is shown in Figure 24-17? _____

_____

_____

25 Animal Diversity

CHAPTER

Directed Reading

Section 25-1 *Sponges, Cnidarians, and Simple Worms* page 479

Sponges

Mark each statement *T* if it is true and *F* if it is false.

_____ **1.** In 1765, zoologists realized that sponges are plants.

_____ **2.** The sponge body is perforated with holes that lead to an inner water chamber.

_____ **3.** Spongin or spicules form the skeleton of a sponge.

_____ **4.** Sponges cannot reproduce sexually.

Cnidarians: Jellyfish and Relatives

Choose the statement from Column B that best matches the term in Column A, and write the letter in the space provided.

Column A	Column B
_____ **5.** cnidaria	**a.** stinging structures found in the cells and outer body surface of cnidarians
_____ **6.** nematocysts	**b.** name comes from the Greek word meaning "nettle"
_____ **7.** polyps	**c.** body form of cnidarians illustrated in Figure 25-3a
_____ **8.** medusae	**d.** body form of cnidarians illustrated in Figure 25-3b

Flatworms

Study the following stages in the life cycle of Schistosoma to determine the order in which they occur. Because there is no beginning to a cycle, begin with the stage marked "1." Write the number of each step in order after that in the blanks provided.

__1__ **9.** Larvae invade freshwater snails.

_____ **10.** Larval worms come into contact with a human in the contaminated water.

_____ **11.** Eggs damage or block small blood vessels and cause internal bleeding.

_____ **12.** Eggs in the feces or urine of infected humans contaminate fresh water.

_____ **13.** Larval worms burrow through the human skin and lay eggs.

_____ **14.** Eggs hatch into larvae.

Roundworms

Place a check mark next to each accurate statement in the space provided.

_____ **15.** A spadeful of soil can contain more than 1 million individual nematodes.

_____ **16.** The body of a roundworm is oval in shape when viewed in cross section.

_____ **17.** Roundworms have no separate sexes.

_____ **18.** The nematode in Figure 25-8 infects humans and pigs.

_____ **19.** The condition shown in Figure 25-10 is caused by the nematode *Wuchereria bancrofti.*

◼ Section 25-2 *Mollusks, Annelids, and Arthropods*   page 487

Mollusks

Complete each statement by underlining the correct word in the brackets.

1. [Nephridia/Eustachia] are small tubules used by mollusks to collect wastes and release them to the outside.

2. Slugs and snails are members of the class [Bivalvia/Gastropoda].

3. Gastropods use a feeding organ described in Figure 25-12b called the [radula/foot].

4. The foot of a [bivalve/cephalopod] is used for digging or secreting substances that allow the organism to attach to solid surfaces.

5. Cephalopods are known for their well-developed [digestive/nervous] systems.

Annelids: Segmented Worms

Mark each statement below *T* if it is true and *F* if it is false.

_____ **6.** According to Figure 25-15, segments at the front of the earthworm's body are specialized for burrowing.

_____ **7.** An earthworm can eat 20 times its own weight in soil in a single day.

_____ **8.** The aquatic leech in Figure 25-16 is a parasitic annelid.

_____ **9.** Modern physicians reject the use of blood-sucking leeches for healing purposes.

(continued on the next page)

Arthropods: The Most Abundant Animals

Complete each statement by writing the correct word in each space provided.

10. There are more than _____ species of arthropods.

11. Arthropods have _____ bodies and are covered with

a(n) _____ of chitin.

12. Arthropods such as _____, _____, and

_____ are sources of food.

▌Section 25-3 *Echinoderms and Chordates*   page 493

Echinoderms: Sea Stars and Their Relatives

Complete each statement by underlining the correct word or phrase in the brackets.

1. As seen in Figure 25-20, all echinoderms show a [five-part/six-part] radial symmetry.

2. The [water vascular system/digestive system] of an echinoderm aids in movement, feeding, and distribution of gases.

Chordates

Read the questions, and write your answer in the space provided.

3. List the four features all chordates share. _____

_____

_____

4. What is the outer body of the tunicate called? _____

_____

_____

5. Which of the four chordate features do adult tunicates retain? _____

_____

_____

(continued on the next page)

6. Name the lancelet shown in Figure 25-22. _____

_____

_____

7. The success of vertebrates is attributed to what features? _____

_____

_____

_____

8. What distinguishing feature gives vertebrates their name? _____

_____

_____

26 Arthropods

CHAPTER

Directed Reading

■ **Section 26-1:** *Spiders and Their Relatives*   page 501

Characteristics of Arachnids

Choose the statement from Column B that best matches the term in Column A, and write the letter in the space provided.

Column A	Column B
_____ **1.** chelicerae	**a.** pair of appendages used to capture and manipulate prey and for courtship and mating
_____ **2.** pedipalps	**b.** members of the class Arachnida
_____ **3.** arachnids	**c.** pair of poison-delivering appendages located at the front of a spider's body

Spiders

Complete each statement by writing the correct word in each space provided.

4. Spiders are believed to be among the first arthropods to successfully live

on _____ .

5. Spiders use _____ _____ to breathe.

6. The silk used to build a spider's web flows from small nozzlelike structures at the end

of a spider's abdomen called _____ .

7. A spider's _____ is used to ensnare prey.

8. The _____ rituals of spiders ensure successful mating.

Other Arachnids

Mark each statement *T* if it is true and *F* if it is false.

_____ **9.** As shown in Figure 26-5, the scorpion's stinger is at the tip of its abdomen.

_____ **10.** Most ticks are free-living on land or in water.

_____ **11.** Mites parasitize almost all groups of animals and many plants.

Horseshoe Crabs

Place a check mark next to each accurate statement.

_____ **12.** Horseshoe crabs have changed substantially in the last 200 million years.

_____ **13.** The five species of horseshoe crabs live on the sandy bottoms of shallow oceans.

■Section 26-2 *Insects, Millipedes, and Centipedes* page 506

Insects

Choose the statement from Column B that best matches the term in Column A, and write the letter in the space provided.

Column A	Column B
_____ **1.** mandibles	**a.** type of metamorphosis illustrated in Figure 26-11 in which the immature form does not resemble the adult
_____ **2.** metamorphosis	
_____ **3.** incomplete metamorphosis	**b.** insect that emerges from the egg during incomplete metamorphosis
_____ **4.** nymph	**c.** each of the changes in the life cycle of an insect
_____ **5.** complete metamorphosis	**d.** series of gradual changes into adulthood
	e. insect's jaws
_____ **6.** soldiers	**f.** insects that pollinate the flowers of plants
_____ **7.** sleeping sickness	**g.** members of a termite colony that protect the colony against attack
_____ **8.** pollinators	
	h. disease carried by biting flies in Africa

Millipedes and Centipedes

Mark each statement below *T* if it is true and *F* if it is false.

_____ **9.** Centipedes, millipedes, and insects all share the same fundamental head structure.

_____ **10.** The name millipede means "hundred feet."

_____ **11.** Some tropical centipedes kill and eat animals as large as frogs and lizards.

(continued on the next page)

■ **Section 26-3** *Crustaceans*

Crustaceans Are Successful Aquatic Arthropods

Complete each statement by underlining the correct word or phrase in the brackets.

1. The crayfish in Figure 26-17 has [eight/ten] walking legs.

2. The exoskeletons of crustaceans are [more/less] watertight than those of insects.

3. One reason that crustaceans have not successfully moved onto land is the lack of [efficient excretory organs/ appendages for sensing the environment].

4. Crustaceans have [gills/book lungs] that are very efficient under water but collapse in air.

Crustacean Diversity

Place a check mark next to each accurate statement.

_____ **5.** The crustacean in Figure 26-18a is a copepod.

_____ **6.** Adult barnacles live inside their calcium shells and float about freely in the ocean.

_____ **7.** The crustacean in Figure 26-18e has five pairs of thoracic legs.

_____ **8.** The water fleas shown in Figure 26-19 are a vital link between producers and the rest of the freshwater food web.

_____ **9.** The tiny crustaceans in Figure 26-20 have no apparent value in the food web of the oceans.

_____ **10.** Our sources of food from the sea would vanish without the copepods.

27 Fishes and Amphibians

CHAPTER

Directed Reading

▌Section 27-1 *Jawless Fishes and Cartilaginous Fishes* page 523

The First Fishes: Class Agnatha

On the line at the left, write the letter of the answer that best completes each sentence.

_____ **1.** The first vertebrates were
  a. jawed fishes.
  b. jawless fishes.
  c. sharks.
  d. salamanders.

_____ **2.** Lampreys and hagfishes are examples of agnathans that
  a. are living today.
  b. have just become extinct.
  c. are on the verge of becoming extinct.
  d. became extinct 360 million years ago.

_____ **3.** Agnathans do not have
  a. a notochord.
  b. cartilage in their skeletons.
  c. a well-developed vertebral column.
  d. gills.

_____ **4.** The agnathan shown in Figure 27-2 is a
  a. placoderm.
  b. Drepanaspis.
  c. jawless hagfish.
  d. parasitic sea lamprey.

Evolution of Jaws

Complete each statement by writing the correct word in each space provided.

5. Figure 27-4 shows how _____ evolved in fishes.

6. Animals with jaws can take advantage of a much more extensive range

of _____ than can jawless animals.

7. The earliest jawed fishes were the _____ .

8. Perhaps because of their jaws and paired fins, placoderms and acanthodians largely

replaced the _____ .

Sharks and Rays: Class Chondrichthyes

Mark each statement *T* if it is true and *F* if it is false.

_____ **9.** The skeletons of sharks and rays are composed of cartilage.

_____ **10.** Figure 27-6b shows the similarity of a shark's scales to its teeth.

_____ **11.** Sharks and rays lack a lateral line system that could enable them to detect animals swimming near them.

_____ **12.** All skates, rays, and sharks have internal fertilization and lay eggs.

■ Section 27-2 *Bony Fishes*   page 528

Structure of a Bony Fish

Place a check mark next to each accurate statement.

_____ **1.** Bony fishes belong to the class Osteichthyes.

_____ **2.** Figure 27-7c shows how the operculum protects the gills of bony fishes.

_____ **3.** A striking similarity between bony fishes and sharks is the presence of a swim bladder.

_____ **4.** Bony fishes detect pressure changes in the water around them through their lateral line system.

_____ **5.** In most bony fishes, fertilization takes place inside the female's body.

Major Groups of Bony Fishes

Choose the statement from Column B that best matches the term in Column A, and write the letter in the space provided.

Column A	Column B
_____ **6.** ray-finned fishes	**a.** species of lobe-finned fishes that was thought to be extinct until 1938
_____ **7.** lobe-finned fishes	**b.** extinct class of fishes that had jaws, paired fins, and bony armor
_____ **8.** coelacanth	**c.** bony fishes whose fins are fan-shaped and are supported by thin bony rays
_____ **9.** Agnatha	
_____ **10.** Acanthodii	**d.** class of fishes that includes ray-finned fishes and lobe-finned fishes
_____ **11.** Placodermi	**e.** bony fishes whose fins are fleshy; extinct forms are ancestors of amphibians
_____ **12.** Chondrichthyes	
_____ **13.** Osteichthyes	**f.** lampreys and hagfishes are members of this class of fishes
	g. extinct class of fishes that had jaws and spiny, paired fins
	h. class of fishes whose members are characterized by jaws, paired fins, and skeletons of cartilage

■ Section 27-3 *Amphibians*   page 533

The First Land Vertebrates

Complete each statement by underlining the correct word or phrase in the brackets.

1. Amphibians were the first vertebrates to [live in the oceans/walk on land].

2. All amphibians except [caecilians/salamanders] have legs.

3. Most amphibians have [gills/lungs] for gas exchange.

4. Because the skin of an amphibian is not resistant to drying, these animals must live in a [moist/dry] environment.

5. Because of their double-loop [digestive/circulatory] system, amphibians can pump blood to their bodies at faster flow rates than can fishes.

The Tie to Water

Study the following steps in the life cycle of a frog to determine the order in which they take place. Then write the number of each step in the blank provided.

_____ **6.** The tadpole's tail and gills recede, while its lungs and front and hind limbs grow.

_____ **7.** Tadpoles change into carnivorous adults that are able to breathe air.

_____ **8.** Fertilized eggs are laid in a wet or moist environment.

_____ **9.** Herbivorous tadpoles emerge from the eggs.

Kinds of Amphibians

Compare the orders of amphibians using the table below.

	Order	Limbs	Tail	Distribution	Example
10.	Gymnophiona				
11.	Anura				
12.	Urodela				

28 Reptiles, Birds, and Mammals

CHAPTER

Directed Reading

▋Section 28-1 *Reptiles*

Reptilian Adaptations to Terrestrial Life

On the line at the left, write the letter of the answer that best completes each sentence.

_____ **1.** The skin of reptiles forms a barrier to water loss because it consists of
   a. skin covered with fur.
   b. scales made of bone.
   c. scales made of lipids and keratin.
   d. feathers.

_____ **2.** Reptiles do not have to travel to water to reproduce because reptilian eggs
   a. can absorb water vapor from the atmosphere.
   b. contain their own water supply.
   c. remain in the body of the female.
   d. receive water from the father.

_____ **3.** The heart of most reptiles, shown in Figure 28-2b, differs from the frog heart shown in Figure 28-2a because it has
   a. a partial division of the ventricle.
   b. a ventricle that is divided into two separate chambers.
   c. a single, undivided ventricle.
   d. no ventricle, only an atrium.

_____ **4.** Which of the following is not an ectotherm?
   a. Fishes
   b. Amphibians
   c. Reptiles
   d. Birds

The Age of Reptiles

Describe dinosaurs and other early reptile groups by completing the following table.

	Group	Habitat	Characteristics
5.	Ichthyosaurs		
6.	Plesiosaurs		
7.	Apatosaurus		
8.	Tyrannosaurus rex		
9.	Pterosaurs		

The Survivors

Mark each statement *T* if it is true and *F* if it is false.

_____ **10.** When a mass extinction occurred about 65 million years ago, only four orders of reptiles were spared: Squamata, Testudines, Crocodylia, and Rynchocephalia.

_____ **11.** Alligators and crocodiles are the reptiles most closely related to dinosaurs.

_____ **12.** Turtles such as the one shown in Figure 28-10 have sharp teeth in their beaks.

_____ **13.** Turtles spend some time in water, while tortoises are almost completely terrestrial.

_____ **14.** Both remaining species of tuataras are found only in New Zealand.

_____ **15.** The snake in Figure 28-12 can devour prey larger than itself because it is jawless.

_____ **16.** Most species of lizards are herbivorous.

_____ **17.** According to Table 28-1, there are ten extinct orders of reptiles.

▮Section 28-2 *Birds*

page 552

Birds Evolved From Reptiles

Read each question, and write your answer in the space provided.

1. What reptilian characteristics do modern birds retain? _____

_____

_____

2. Describe the two types of bird feathers as shown in Figure 28-15. _____

_____

_____

3. What role does the sternum play in a bird's ability to fly? _____

_____

_____

4. Describe a bird's circulatory system. Why is the design of this system important

in terms of a bird's ability to fly? _____

_____

_____

How Birds Fly

Mark each statement below *T* if it is true and *F* if it is false.

_____ **5.** A bird creates the pressure difference needed to produce lift through the arch of its wing or by changing its angle of attack.

_____ **6.** The thrust birds need to move forward is created by the upstroke of the wings.

_____ **7.** The most common method of flight is called flapping flight.

_____ **8.** Soaring requires a great deal of energy and allows birds to remain aloft for only very short periods of time.

Major Orders of Birds

Complete each statement by writing the correct word in each space provided.

9. Hummingbirds are members of the order _____ .

10. A bird with well-developed vocal organs is the _____ .

11. An example of a member of the order Falconiformes is the _____ .

12. An owl's large _____ are particularly useful because it hunts in the dark.

13. A _____ is an example of Spheniciforme.

▌Section 28-3 *Introduction to Mammals*   page 558

Origin of Mammals

Read each question, and write your answer in the space provided.

1. What characteristic distinguishes the first mammals from their reptilian ancestors, the

therapsids? _____

_____

2. What has happened to mammals since the dinosaurs disappeared 65 million years ago?

_____

_____

3. Tell how climate changes during the past 15 million years have affected mammals.

_____

_____

Mammalian Characteristics

Mark each statement below *T* if it is true and *F* if it is false.

_____ **4.** Mammals are characterized by the presence of hair or fur on the body and the ability to feed their young with milk from the mammary glands.

_____ **5.** Mammals are ectothermic and are thus unable to inhabit cold regions.

_____ **6.** There is no mixing of deoxygenated and oxygenated blood in mammalian hearts.

Egg-Laying Mammals: The Monotremes

Complete each statement by writing the correct word in each space provided.

7. _____ are the only group of mammals that lay eggs.

8. The _____ and the _____ shown in Figure 28-19 represent the only remaining species of monotremes.

9. Like all other mammals, monotremes have _____ and

_____ glands.

Pouched Mammals: The Marsupials

Place a check mark next to each accurate statement.

_____ **10.** The primary difference between marsupials and placental mammals is their differing patterns of embryonic development.

_____ **11.** The newborn marsupial crawls into the pouch of its mother where it nurses and grows.

_____ **12.** All species of marsupials are found on every continent in the world.

(continued on the next page)

True Placental Mammals

Choose the phrase from Column B that best matches the term listed in Column A, and write your answer in the space provided.

	Column A		Column B
_____	**13.** Rodentia	**a.**	land-living predators including dogs, bears, cats, and wolves
_____	**14.** Chiroptera	**b.**	rodentlike with hind legs adapted for jumping
_____	**15.** Insectivora	**c.**	includes squirrels, rats, and beavers
_____	**16.** Carnivora	**d.**	largest living land animals; have long trunks and tusks
_____	**17.** Primates	**e.**	marine carnivores such as seals and walruses
_____	**18.** Artiodactyla	**f.**	the only flying mammals
_____	**19.** Cetacea	**g.**	includes humans and apes
_____	**20.** Lagomorpha	**h.**	hoofed mammals with one or three toes
_____	**21.** Pinnipedia	**i.**	the most primitive placental mammals
_____	**22.** Edentata	**j.**	aquatic mammals with streamlined bodies; no hind limbs
_____	**23.** Perissodactyla	**k.**	large, hoofed herbivores with two or four toes
_____	**24.** Proboscidea	**l.**	includes sloths, anteaters, and armadillos

■ Section 28-4 *Mammalian Adaptations*   page 564

Hair Has Many Functions

Complete each statement by underlining the correct word or phrase in the brackets.

1. A mammal's coat of hair is made up of [guard hair and underhair/scales and bones].

2. A primary function of a mammal's hair is to [insulate it against heat loss/protect it from the sun's rays].

3. Cats and dogs use their whiskers as [camouflage/sensory structures].

Claws, Hooves, Horns, and Antlers

Mark each statement below *T* if it is true and *F* if it is false.

_____ **4.** Keratin is an important component of claws, hooves, horns, and antlers.

_____ **5.** The horns of the South African springbok shown in Figure 28-23a are shed each year.

_____ **6.** The antlers of the mule deer in Figure 28-23b are used for combat and to attract females.

Food and Feeding

Read each question, and write your answer in the space provided.

7. How does Figure 28-24 show the distinction between the teeth of carnivores and those

of herbivores? _____

_____

8. What mutualistic partners assist herbivores, such as cows, antelopes, and deer, to

digest cellulose? _____

_____

9. Why must herbivores consume a greater quantity of food than carnivores? _____

_____

Flying Mammals

Place a check mark next to each accurate statement.

_____ **10.** Bats are the only mammals capable of powered flight.

_____ **11.** Bats have wings covered with feathers.

_____ **12.** Bats use a sonar system to navigate and to capture prey.

29 The Human Body

Directed Reading

■ **Section 29-1** *An Inside View of the Human Body*   page 575

Similar Cells Form Tissues

Choose the statement from Column B that best matches the term in Column A, and write the letter in the space provided.

Column A

_____ **1.** tissue

_____ **2.** epithelial tissue

_____ **3.** exocrine gland

_____ **4.** muscle tissue

_____ **5.** nerve tissue

_____ **6.** connective tissue

Column B

a. tissue with three distinct types: defensive, structural, and sequestering

b. tissue made of cells that contract or relax in response to electrical signals

c. examples of this tissue include skin and the inner and outer covering of the internal organs

d. tissue found in the brain, nerves, and sense organs

e. cells or group of cells that produce and release secretions onto the body's surface

f. group of similar cells that work together to perform a specific function

Tissues Form Organs

Complete each statement by writing the correct word in each space provided.

7. A structure made of a collection of tissues that work together to perform a specific function in the body is called a(n) _____ .

8. The largest organ of the body, the _____, is composed of all four types of _____ .

9. Organs that work together form a(n) _____ _____ .

10. The organ systems highlighted in Figure 29-3 are the _____ system, the _____ system, and the _____ system.

▌Section 29-2 *Skin*

The Dermis

Choose the statement from Column B that best matches the term in Column A, and write the letter in the space provided.

Column A	Column B
_____ **1.** epidermis	**a.** help to keep the body cool through the evaporation of moisture from the skin
_____ **2.** dermis	**b.** inner part of the skin
_____ **3.** nerves	**c.** carry nutrients to the skin and waste products away
_____ **4.** muscles	**d.** part of the dermis that enables us to sense pressure, temperature, and pain
_____ **5.** blood vessels	**e.** outer layer of skin
_____ **6.** sweat glands	**f.** contract and pull hairs in the skin upright when one is cold or afraid

The Epidermis

Mark each statement below *T* if it is true and *F* if it is false.

_____ **7.** The outermost layer of the epidermis is made of layers of live cells.

_____ **8.** Keratin and the oil glands in the dermis work together to make skin water-proof and prevent the loss of water by evaporation.

_____ **9.** Figure 29-6c shows that in nails new cells are formed in the white half-moons.

_____ **10.** New epidermal cells produced in the epithelium are called basal cells.

_____ **11.** Melanin is not present in black skin, but is found in large quantities in the skin of albinos.

Skin Disorders

Complete each statement by underlining the correct word or phrase in the brackets.

12. Acne is produced by [oil/sweat] glands, [viruses/bacteria], and [basal cells/cellular debris].

13. Prolonged exposure to sun rays can lead to a [more rapid/slower] aging of the skin and [an overabundance of vitamin D/skin cancer].

14. Seventy-five percent of the deaths attributed to skin cancer are caused by [malignant melanomas/basal cell cancer].

15. Damage to the [oil glands/p53 gene] may allow skin cancer to develop.

▎Section 29-3 *Bones*

Bone Structure and Growth

On the line at the left, write the letter of the answer that best completes each sentence

_____ 1. The four minerals in the bone cells that give bones their strength are
- **a.** iron, carbon, phosphorus, and calcium.
- **b.** carbon dioxide, calcium, phosphorus, and manganese.
- **c.** calcium, phosphorus, magnesium, and manganese.
- **d.** oxygen, carbon, calcium, and magnesium.

_____ 2. The spaces between the bone cells and minerals inside the ends of a long bone are filled with
- **a.** cartilage.
- **b.** minerals.
- **c.** blood.
- **d.** marrow.

_____ 3. Because of the risk of infection, the most serious type of broken bone is a
- **a.** simple fracture.
- **b.** compound fracture.
- **c.** sprain.
- **d.** crack.

_____ 4. The condition that results from bones becoming less dense and more brittle is
- **a.** osteoporosis.
- **b.** a compound fracture.
- **c.** tendinitis.
- **d.** cardiovascular disease.

_____ 5. Which of the following does not help prevent osteoporosis?
- **a.** Regular exercise
- **b.** Nutritious diet
- **c.** Protection from ultraviolet radiation
- **d.** Adequate levels of sex hormones

(continued on the next page)

The Skeleton

Choose the statement from Column B that best matches the term in Column A, and write the letter in the space provided.

	Column A		Column B

       6. skull   **a.** framework of bone to which the legs are attached

       7. spine   **b.** framework of bone to which the arms are attached

       8. ribs   **c.** injury that results when ligaments are stretched too far

       9. shoulder girdle   **d.** place where two or more bones connect

      10. pelvic girdle   **e.** fused bones that protect the brain and form the shape of the face

      11. hipbones

      12. joint   **f.** made up of vertebrae that support the trunk

      13. ligaments   **g.** connective tissue that joins bones together

      14. sprain   **h.** attached to the sternum by cartilage; protects the heart, lungs, and some other organs

                                     i. part of the pelvic girdle directly attached to the spine

■ Section 29-4 Muscles

page 590

The Actions of Muscles

Compare the three main types of human muscle tissue by completing the following table.

	Muscle	Location	Function
1.	Skeletal		
2.	Smooth		
3.	Cardiac		

Making Your Skeleton Move

Place a check mark next to each accurate statement.

      4. Two sets of muscles are required to move a bone.

      5. If both sets of muscles were to contract at the same time, the bone would not move.

      6. Most muscles are attached directly to the bone.

      7. Figure 29-19b shows how the threads of actin and myosin slide past each other and the entire muscle shortens.

      8. Whether you are picking up a pencil or a bowling ball, the same number of muscle cells contract.

_____ **9.** The constant complete contraction of your muscles is known as muscle tone.

What Exercise Does for Muscles

Explain the difference between each set of terms in the space provided.

10. aerobic exercise/anaerobic exercise

_____

_____

_____

11. fast-twitch muscles/slow-twitch muscles

_____

_____

_____

12. anabolic steroids/testosterone

_____

_____

_____

13. muscle strain/tendinitis

_____

_____

_____

30 The Nervous System

Directed Reading

■ **Section 30-1** *Nerve Impulses*   page 601

Structure of a Neuron

Choose the statement from Column B that best matches the term in Column A, and write the letter in the space provided.

Column A	Column B
_____ **1.** neurons	**a.** short, branched extensions of a neuron that allow it to receive input from other cells
_____ **2.** nerves	**b.** nerve cells specialized for conducting information
_____ **3.** dendrites	**c.** long fibers that enable neurons to transmit information to other cells
_____ **4.** axons	**d.** bundles of neurons in the form of thin cables

Transmission of Nerve Impulses

Complete each statement by writing the correct word in each space provided.

5. The unequal concentration of sodium and potassium ions results in a difference in

voltage across the neuron membrane called the _____

_____ .

6. The sudden reversal of electrical charge across the neuron membrane is called

a(n) _____ _____ .

7. A(n) _____ _____ is a sequence of action
potentials progressing along a neuron.

Speeding Up Nerve Impulses

Mark each statement *T* if it is true and *F* if it is false.

_____ **8.** A myelin sheath insulates a neuron's axons.

_____ **9.** Figure 30-3 shows how a nerve impulse "jumps" from node to node
on an axon.

_____ **10.** Sleeping sickness destroys large patches of myelin around the neurons in the
brain and spinal cord.

Transmission Across Synapses

Place a check mark next to each accurate statement.

_____ **11.** All neurons touch each other directly.

_____ **12.** A single axon may form synapses with many other neurons.

_____ **13.** A nerve impulse is carried across the synapse by neurotransmitters.

_____ **14.** All neurotransmitters excite nerve impulses across a synapse.

Drugs and the Nervous System

Complete each statement by underlining the correct word or phrase in the brackets.

15. Psychoactive drugs are addictive because they alter tissues in the [circulatory/nervous] system.

16. In Figure 30-6, Prozac is shown to work as an antidepressant because it blocks reabsorption of [serotonin/melatonin] from the synapse.

17. Figure 30-7 illustrates how the change in the number of [sodium-potassium pumps/receptors] results in addiction.

18. The psychoactive drug cocaine obstructs the reabsorption of [dopamine/enkaphalins], the neurotransmitter that transmits pleasure messages to the brain.

19. Alcohol affects normal brain functioning by altering the structure of the [neurotransmitters/cell membrane], producing changes in the shape of receptors.

▌Section 30-2 *The Nervous System*
page 611

Organization of the Human Nervous System

Place a check mark next to each accurate statement in the space provided.

_____ **1.** Figure 30-9 shows that the human nervous system is organized into the central nervous system and the peripheral nervous system.

_____ **2.** The central nervous system is composed of the brain, the spinal cord, sensory neurons, and motor neurons.

_____ **3.** The organs of the peripheral nervous system are protected by the vertebrae and the skull.

The Central Nervous System

Choose the statement from Column B that best matches the term in Column A, and write the letter in the space provided.

Column A	Column B
_____ **4.** cerebrum	**a.** regulates such vital body processes as heartbeat, respiration, and blood pressure
_____ **5.** cerebellum	
_____ **6.** brain stem	**b.** center of intellect, memory, language, and consciousness
_____ **7.** hypothalamus	**c.** controls the body's homeostasis and regulates hunger
	d. responsible for well-coordinated body movements

Mapping the Brain

Complete each statement by underlining the correct word or phrase in the brackets.

8. [Language expression/Learning] takes place when two or more pieces of information are linked into a pattern of connections between neurons.

9. Language processing is very complex and takes place in several regions of the [left/right] hemisphere of the brain.

10. The technology for studying the brain that is shown in Figure 30-12a is [magnetic resonance imaging/CAT scans].

11. [CAT/PET] scans can be used to diagnose mental illnesses such as Alzheimer's disease and schizophrenia.

Peripheral Nervous System

Write the term that is being described in each space provided.

_____ **12.** all of the nervous system outside the spinal cord and brain

_____ **13.** carry nerve impulses from sense organs to the central nervous system

_____ **14.** carry information from the central nervous system to a muscle or gland

_____ **15.** sudden, involuntary movement, such as the blinking of an eyelid to protect the eye

_____ **16.** carries messages to muscles and glands that typically work without our noticing

■ Section 30-3 *The Sense Organs* page 621

Sensory Receptors

Read each question and write your answer in the space provided.

1. What sensory receptors are found in the skin and to what stimuli do they respond?

_____

_____

2. Where are taste bud cells located? _____

_____

3. How do rod and cone cells process light? _____

_____

Receptors in the Ear

Mark each statement below *T* if it is true and *F* if it is false.

_____ **4.** The ear can detect sound waves and establish equilibrium.

_____ **5.** The receptor cells for hearing are located all along the fluid-filled canals and chambers in your inner ear, as shown in Figure 30-15b.

_____ **6.** The hair cells of the cochlea can be permanently damaged by strong vibrations from loud music.

Receptors in the Eye

Complete each statement by underlining the correct word or phrase in the brackets.

7. There are [two/three] types of photoreceptors in your eyes.

8. Cones are [able/not able] to detect color.

9. Figure 30-16c shows how each eye receives [all/about three-quarters] of an image.

10. Humans [are/are not] the only animals that can see color.

Receptors for Taste and Smell

Place a check mark next to each accurate statement.

_____ **11.** Receptor cells for taste are located in the taste buds that line the upper surface of the tongue.

_____ **12.** Humans can detect four basic tastes: sweet, sour, salty, and bitter.

_____ **13.** There are only three types of receptors for smell.

31 Hormones

CHAPTER

Directed Reading

Section 31-1: *What Hormones Do*

page 631

Hormones: Chemical Signals

Mark each statement *T* if it is true and *F* if it is false.

_____ **1.** A hormone is a chemical signal, made in one part of the body and delivered to another, that regulates the body's activities.

_____ **2.** Most of the hormones in the body are produced in the brain.

_____ **3.** The endocrine system regulates metabolism, growth, reproduction, and maintenance of homeostasis.

_____ **4.** The effects of nerve impulses last longer than those of hormones.

The Hypothalamus-Pituitary Connection

Choose the statement from column B that best matches the term in column A, and write the letter in the space provided.

Column A	Column B
_____ **5.** pituitary gland	**a.** causes glands in the mother's breasts to produce milk
_____ **6.** growth hormone	**b.** causes the kidneys to form more concentrated urine, conserving water in the body
_____ **7.** prolactin	**c.** aids in the release of milk from the mother's breasts and also causes contractions of the uterus during and shortly after labor
_____ **8.** oxytocin	**d.** stimulates growth of the body, particularly the skeleton
_____ **9.** vasopressin	**e.** once called the "master gland" because many of its hormones regulate other endocrine functions

Regulating Hormone Release

Complete each statement by writing the correct word in each space provided.

10. Upon receiving _____, the endocrine system adjusts the amount of hormones being made or released.

11. _____ feedback stimulates the output of more hormone, while

_____ feedback inhibits the production of more hormone.

12. Figure 31-4 shows how hormones regulate _____

_____ by controlling the body's rate of metabolism.

13. The hormone _____ lowers the level of calcium in the blood, and

_____ _____ raises it.

■ Section 31-2 *How Hormones Work*   page 636

Hormone Receptor Proteins

Read each question, and write your answer in the space provided.

1. What is taking place in Figure 31-6? _____

_____

_____

2. Where are receptor proteins for hormones located in the cell? _____

_____

_____

Steroid Hormones

Mark each statement below *T* if it is true and *F* if it is false.

_____ **3.** Steroid hormones include the sex hormones and adrenal cortex hormones.

_____ **4.** Receptors for most of the sex hormones are located in the nucleus of the cell.

_____ **5.** Steroid hormones lack the ability to switch genes off and on.

_____ **6.** Figure 31-7 shows how steroid hormones influence the development of male sexual traits.

Peptide Hormones

Complete each statement by underlining the correct word or phrase in the brackets.

7. [Peptide/Steroid] hormones are made of chains of amino acids.

8. Peptide hormones bind to receptors on the [cell membrane/nuclear membrane].

9. Second messengers such as [glucagon/cyclic AMP] activate enzymes in the cytoplasm of target cells.

10. One hormone binding to a receptor in the cell's plasma membrane can generate the formation of [many/a few] second messengers in the cytoplasm.

▌Section 31-3 *Glands and Their Functions*   page 639

The Adrenal Glands

Complete each statement by writing the correct word in each space provided.

1. The _____ _____ are two almond-sized glands located on top of the kidneys.

2. Each adrenal gland is really two glands in one—the inner _____

_____ and the outer _____ _____.

3. The hormones _____ and _____ work with the autonomic nervous system to produce the "fight or flight" reaction.

4. Figure 31-10 illustrates how long-term stress can endanger the body's

_____ .

The Thyroid Gland

Mark each statement below *T* if it is true and *F* if it is false.

_____ **5.** The thyroid gland regulates the body's metabolic rate by releasing cortisone and other hormones containing iodine.

_____ **6.** Graves' disease results from the production of too much thyroxine by the thyroid gland.

_____ **7.** The goiter in Figure 31-12a is the result of hypothyroidism.

_____ **8.** A diet that contains plenty of calcium can prevent goiter from developing.

Other Glands and Hormones

Choose the statement from column B that best describes the term in column A, and write your answer in the space provided.

Column A	Column B
_____ **9.** testerone	**a.** female sex hormone that influences the maturation of reproductive organs
_____ **10.** estrogen	
_____ **11.** progesterone	**b.** male sex hormone
_____ **12.** thymosin	**c.** elevates calcium levels in blood
_____ **13.** melatonin	**d.** contains light-sensitive cells that mark changes in the length of days and nights
_____ **14.** parathyroid hormone	**e.** stimulates the development of specialized white blood cells called T cells
	f. female sex hormone that stimulates breast development

The Pancreas

Place a check mark next to each accurate statement in the space provided.

_____ **15.** According to the diagram in Figure 31-13, the hormone insulin raises blood sugar levels after meals, while the hormone glucagon lowers blood sugar levels between meals.

_____ **16.** When a person's body does not make enough insulin, insulin-dependent diabetes mellitus develops in childhood or early adolescence.

_____ **17.** Prostaglandins function as localized hormones in the body.

_____ **18.** All of the drugs that inhibit the production and release of prostaglandins cause Reye's syndrome.

32 Circulation and Respiration

CHAPTER

Directed Reading

Section 32-1 *Circulation*

Transporting Materials Through the Body

Place a check mark next to each accurate statement.

_____ **1.** Large animals rely solely on diffusion to supply food, oxygen, and other materials and to carry away wastes.

_____ **2.** The circulatory systems of mammals and birds help maintain their constant body temperatures.

_____ **3.** William Harvey demonstrated that blood circulates from the heart to the rest of the body and back again in a closed circuit.

Blood: A Liquid Tissue

Choose the statement from Column B that best matches the term in Column A, and write the letter in the space provided.

Column A	Column B
_____ **4.** plasma	**a.** cell that carries oxygen from the lungs to all of the cells of the body
_____ **5.** red blood cells	
_____ **6.** hemoglobin	**b.** protein molecule in which oxygen binds easily to the iron
_____ **7.** white blood cells	**c.** colorless and irregularly shaped cells that protect the body against foreign invaders
_____ **8.** platelets	**d.** fragments of large white blood cells that play an important role in the clotting of blood
	e. protein-rich substance that makes blood a liquid

Blood Vessels

Complete the following table about the three types of blood vessels.

	Vessel	Function	Structure
9.	Arteries		
10.	Veins		
11.	Capillaries		

Blood Types

Mark each statement *T* if it is true and *F* if it is false.

_____ **12.** Blood type is determined by the presence or absence of special marker proteins on the surfaces of the nuclear membrane of red blood cells.

_____ **13.** One blood group system uses the letters A, B, AB, and O to label the different blood types.

_____ **14.** Red blood cells that have the Rh cell surface marker on their surface are termed Rh-negative.

The Lymphatic System

Read each question, and write your answer in the space provided.

15. Describe the basic structure and function of the lymphatic system as shown

in Figure 32-5. _____

_____

_____

_____

_____

16. How does the lymphatic system help defend the body against infection? _____

_____

_____

_____

_____

(continued on the next page)

▌Section 32-2 *How Blood Flows*

page 658

The Heart

Choose the statement from Column B that best matches the term in Column A, and write the letter in the space provided.

	Column A		Column B
_____	**1.** atrium	**a.**	transports oxygen-rich blood from lungs into left atrium
_____	**2.** ventricle		
_____	**3.** pacemaker	**b.**	receives oxygen-poor blood from lower body
_____	**4.** pulmonary arteries	**c.**	transports oxygen-rich blood to lower body
_____	**5.** pulmonary veins	**d.**	upper chamber of the heart that receives incoming blood
_____	**6.** superior vena cava		
_____	**7.** inferior vena cava	**e.**	sends electrical signals to cause heart muscle cells to contract
_____	**8.** aorta	**f.**	carries oxygen-rich blood to other arteries
_____	**9.** descending aorta	**g.**	receives oxygen-poor blood from upper body
		h.	transports oxygen-poor blood to the lungs
		i.	lower chamber of the heart that pumps blood out

Circulatory Pathways

Place the sequence of events for tracing the pathway of bloodflow through the circulatory system in order by numbering the steps.

_____ **10.** Oxygen-rich blood is pumped from the left ventricle through the aorta and the descending aorta to the body.

_____ **11.** From the right ventricle, the oxygen-poor blood is pumped through the pulmonary arteries.

_____ **12.** Pulmonary veins transport the oxygen-rich blood from the lungs to the left atrium.

_____ **13.** Oxygen-rich blood is pumped from the left atrium to the left ventricle.

_____ **14.** Oxygen-poor blood enters the right atrium from both the superior and inferior venae cavae.

_____ **15.** Pulmonary arteries transport the oxygen-poor blood to the lungs.

_____ **16.** Oxygen-poor blood is pumped from the right atrium to the right ventricle.

Blood Pressure

Place a check mark next to each accurate statement.

_____ **17.** The force exerted on the walls of the blood vessels when the heart forces blood into the arteries is called blood pressure.

_____ **18.** Blood pressure is expressed in terms of inches of mercury.

_____ **19.** Blood pressure rises with increasing age in industrialized countries.

Blood Chemistry

Complete each statement by underlining the correct word or phrase in the brackets.

20. Extreme imbalances in the concentrations of solids and gases in the blood indicate [disease/good health].

21. An analysis of blood plasma provides information about the body's system of complex interdependences that are required to sustain [adequate oxygen supply/homeostasis].

22. Because [LDL-cholesterol/HDL-cholesterol] is removed from body tissues, it is frequently called "good cholesterol."

Diseases of the Heart and Blood Vessels

Read each question, and write your answer in the space provided.

23. What does Figure 32-12 illustrate? _____

_____

_____

_____

24. What is hypertension? _____

_____

_____

_____

■ Section 32-3 *The Respiratory System*   page 670

Lungs and Breathing

Place the sequence of events for tracing the pathway of air into the respiratory system in order by numbering the steps.

_____ **1.** Air enters the larynx, or voice box, and the trachea, or windpipe.

_____ **2.** Air enters the smaller branches, the bronchioles, and finally the alveoli, the smallest tubes.

_____ **3.** Air enters the body through the mouth or nose.

_____ **4.** Blood in the capillaries of alveoli picks up oxygen and releases carbon dioxide.

_____ **5.** Air enters the lower end of the trachea, which divides into two branches that reach deep into the lungs.

_____ **6.** Air travels to the pharynx, a tube at the back of the nose and mouth.

Gas Exchange

Mark each statement below T if it is true and F if it is false.

_____ **7.** Gas exchange occurs when oxygen in the alveoli diffuses into the blood in the capillaries and carbon dioxide in the blood diffuses into the air of the alveoli.

_____ **8.** The only way that carbon dioxide can be eliminated from cells is by dissolving in the plasma of blood.

Regulation of Breathing

Complete each statement by underlining the correct word or phrase in the brackets.

9. Receptors in the [spinal cord and lymphatic system/brain and circulatory system] enable the body to automatically regulate oxygen and carbon dioxide concentrations by sending signals to the brain.

10. The concentration of carbon dioxide in the blood has a [greater/lesser] effect on the regulation of breathing than does the concentration of oxygen.

Diseases of the Respiratory System

Compare three diseases of the respiratory system by completing the following table.

	Disease	Description
11.	Asthma	
12.	Emphysema	
13.	Lung cancer	

33 The Immune System

CHAPTER

Directed Reading

▌Section 33-1 *First Line of Defense*

Keeping Pathogens Out

Complete each statement by writing the correct word in each space provided.

1. Swift replacement of _____ _____ enables punctures or cuts in the skin to be sealed promptly.

2. The internal surfaces of the natural openings of the skin, such as mouth, nostrils, and

eyes, are protected by _____ _____ .

3. Some _____ found in saliva and tears can destroy bacteria.

Organizing the Defense

Choose the statement from Column B that best matches the term in Column A, and write the letter in the space provided.

Column A	Column B
_____ **4.** leukocytes	**a.** type of white blood cells that specialize in longer, more serious pathogen invasions
_____ **5.** phagocytes	**b.** another name for white blood cells
_____ **6.** lymphocytes	
_____ **7.** inflammatory response	**c.** type of phagocyte with increased defensive power because it has a longer life than other white blood cells
_____ **8.** immune response	**d.** type of white blood cells that work best against short, minor invasions by pathogens
_____ **9.** macrophages	**e.** body's defense mechanism to injury, indicated by redness and swelling
	f. immune system's attack on a specific pathogen

Antigen Recognition

Mark each statement *T* if it is true and *F* if it is false.

_____ **10.** The recognition and display of foreign marker proteins by macrophages trigger the immune response.

_____ **11.** Marker proteins that trigger the immune response are called antibodies.

_____ **12.** Figure 33-5 shows what happens when blood types that are not compatible are mixed.

_____ **13.** Based on the information in Table 33-1, a person with type O blood can only receive transfusions of blood from people with type A blood.

■ Section 33-2 *The Immune Response* page 683

Main Line of Defense

Read each question, and write your answer in the space provided.

1. How is an influenza virus most likely to enter your body?

_____

_____

_____

_____

2. What is the immune response?

_____

_____

_____

_____

T Cells: Command and Attack

Mark each statement below *T* if it is true and *F* if it is false.

_____ **3.** Helper T cells direct the activities of killer T cells and suppressor T cells.

_____ **4.** Figure 33-7 shows how macrophages divide rapidly to begin the immune response.

_____ **5.** The human body can respond to millions of different antigens because it manufactures millions of different types of T cells.

B Cells: Biological Warfare

Complete each statement by underlining the correct word or phrase in the brackets.

6. As illustrated in Figure 33-8a, helper T cells and B cells bind to the [macrophages/viruses] displaying the antigens they recognize.

7. Figure 33-8b shows how B cells divide and develop into [T cells/plasma cells], which then release thousands of antibodies per second.

8. Figures 33-8c and 33-8e depict how antibodies [bind to/destroy] infected cells and free virus particles.

Pacing the Immune System

Place a check mark next to each accurate statement.

_____ **9.** Moderate fevers discourage the growth of pathogens and stimulate the action of macrophages.

_____ **10.** Two or more weeks into the immune response, the large number of supressor T cells encourages the growth of even larger numbers of B and T killer cells.

_____ **11.** An immune response to a previously encountered pathogen is known as a secondary immune response.

_____ **12.** All vaccines produce lifetime immunity.

■ Section 33-3 *Immune System Failure*   page 688

Immune Overreaction

Complete each statement by underlining the correct word or phrase in the brackets.

1. An allergy is an immune response against a [pathogenic/nonpathogenic] antigen.

2. The immune system of people with [autoimmune diseases/allergies] loses its ability to distinguish self from nonself.

3. The immune system of people with [Graves' disease/multiple sclerosis] attacks and destroys the myelin sheath that covers motor nerves.

Cancer: Unrestrained Cell Division

Mark each statement below *T* if it is true and *F* if it is false.

_____ **4.** A major function of the immune system is to protect against cancer.

_____ **5.** Mutation in the cells of the body is responsible for all cancerous cells.

_____ **6.** Figure 33-13 shows how a killer T cell destroys a cancer cell.

AIDS: Immune System Collapse

Complete each statement by writing the correct word in each space provided.

7. HIV cripples the immune system by invading _____ and

_____ _____ cells and transforming these cells into virus factories.

8. You can only get HIV by contact with _____ _____or

_____ _____ of people infected with HIV.

9. If an expectant mother is infected with HIV, her child has a _____ chance of being infected during pregnancy.

AIDS Is a Worldwide Disease

Place a check mark next to each accurate statement.

_____ **10.** It is estimated that there will be 30-40 million cases of AIDS by the year 2000.

_____ **11.** AIDS is currently one of the two biggest killers of women in Africa.

_____ **12.** The number of AIDS cases is increasing in industrialized countries.

34 Digestion and Excretion | Directed Reading

■ Section 34-1 *Nutrition: What You Eat and Why* page 699

Nutrition and Health

Read each question, and write your answer in the space provided.

1. What are the two major factors that lead to nutritional problems in the United States?

_____

_____

2. What six essential nutrients are required for life?

_____

_____

Food Energy

Choose the statement from Column B that best matches the term in Column A, and write the letter in the space provided.

Column A	Column B
_____ **3.** calorie	**a.** fats found in animal fats and animal protein that are linked to cardiovascular disease, obesity, and cancer
_____ **4.** complex carbohydrates	**b.** also called starches, examples include cereals, grains, and beans
_____ **5.** saturated fats	
_____ **6.** monounsaturated fats	**c.** name given a food's protein if the food contains all the essential amino acids
_____ **7.** polyunsaturated fats	**d.** amino acids that the body cannot make
_____ **8.** essential amino acids	**e.** fats found in safflower, corn, and soybean oil
_____ **9.** "complete" protein	**f.** fats found in canola, olive, and peanut oils
	g. unit of measurement used to describe the amount of energy provided by the energy nutrients

Regulatory Nutrients

Complete each statement by writing the correct word in each space provided.

10. The most vital nutrient is _____ .

11. Plants absorb _____ from the water or soil, and animals get these nutrients by eating plants or plant-eating animals.

12. _____ are needed by humans to help activate enzymes and regulate the release of energy in the body.

Eating Disorders

In the table below, contrast the two eating disorders anorexia nervosa and bulimia and the damage that can be done to the body by each.

	Eating disorder	Behavior	Resulting damage
13.	Anorexia nervosa		
14.	Bulimia		

■ Section 34-2 *The Digestive System*   page 706

Digestion

Complete each statement by underlining the correct word or phrase in the brackets.

1. Saliva in the mouth contains [digestive enzymes/bacteria] that begin the process of digestion.

2. As illustrated in Figure 34-5a, food moves from the mouth into a muscular tube called the [trachea/esophagus].

3. Food moves down the esophagus with the aid of mucus and [agitation/peristalsis].

4. Hydrochloric acid and [pepsin/glucagon] break down proteins into small chains of amino acids.

(continued on the next page)

Activity in the Intestines

Choose the statement from Column B that best matches the term in Column A, and write the letter in the space provided.

	Column A		Column B

      Column A

_____ **5.** small intestine

_____ **6.** duodenum

_____ **7.** villi

_____ **8.** microvilli

_____ **9.** liver

_____ **10.** pancreas

_____ **11.** large intestine

      Column B

a. organ that secretes bile so that fat globules can be broken down and absorbed

b. projections in the intestinal lining

c. organ located in the lower abdomen that absorbs nutrients released by digestion

d. organ that stores, compacts, and eliminates undigestible material

e. organ that secretes digestive enzymes that break down carbohydrates, proteins, and fats

f. hairlike projections on each villus

g. entrance to the small intestine where the rest of digestion takes place

▌Section 34-3 The Excretory System

page 710

Kidney Form and Function

On the line at the left, write the letter of the answer that best completes each sentence.

_____ **1.** Ammonia and other nitrogen-containing wastes are eliminated from the human body in the form of
    a. feces.
    b. uric acid.
    c. urea.
    d. purine.

_____ **2.** The system illustrated in Figure 34-8 is the
    a. excretory system.
    b. digestive system.
    c. reproductive system.
    d. circulatory system.

_____ **3.** Each human kidney contains over 1 million _____ that serve as blood-cleaning units.
    a. stones
    b. microvillus
    c. enzymes
    d. nephrons

Urine Formation

Mark each statement below _T_ if it is true and _F_ if it is false.

_____ **4.** The first stage of urine formation is called titration.

_____ **5.** About 1 L of urine is produced each day.

_____ **6.** During reabsorption, water and solutes move out of the nephron and are sent back into the bloodstream.

_____ **7.** Urine exits the body through the ureter, a tube that leads outside the body.

Kidney Disorders and Treatment

Complete each statement by writing the correct word in each space provided.

8. The common kidney disorder _____ _____ are deposits of uric acid, calcium salts, and other substances.

9. Kidney stones may be eliminated naturally from the body or be removed by

_____ procedures.

10. Patients with kidney disorders may have their blood artificially filtered by

_____ .

35 Reproduction and Development

Directed Reading

■ Section 35-1 *The Male Reproductive System*

Structure of Sperm

1. In the space below, sketch a sperm and show the location of the following: head, mitochondria, tail, digestive enzymes.

The Path Traveled by Sperm

Trace the pathway of sperm from production to ejaculation by placing the following events in order. Then write the number of each step in the blank provided.

_____ **2.** Sperm travel through the urethra and out of the penis.

_____ **3.** Sperm are stored temporarily in the epididymis.

_____ **4.** Sperm travel through the vas deferens to the urethra.

_____ **5.** Sperm are produced in the testes.

Male Hormones and Reproduction

Mark each statement *T* if it is true and *F* if it is false.

_____ **6.** The male hormone testosterone is produced in the testes.

_____ **7.** Testosterone production begins at adolescence.

_____ **8.** Testosterone causes an embryo to develop a penis and scrotum.

_____ **9.** Production of sperm and testosterone are regulated by two pituitary hormones: LH and FSH.

_____ **10.** A positive-feedback system maintains a constant level of testosterone in the blood.

■ Section 35-2 *The Female Reproductive System* page 723

Structure of the Female Reproductive System

Choose the statement from Column B that best matches the term in Column A, and write the letter in the space provided.

Column A	Column B
_____ **1.** ova	**a.** organs responsible for producing eggs
_____ **2.** ovaries	**b.** carries the egg from an ovary into the uterus
_____ **3.** fallopian tube	**c.** muscular tube that connects the uterus to the outside of the body
_____ **4.** uterus	
_____ **5.** cervix	**d.** tubular ring of strong muscles around the lower part of the uterus
_____ **6.** vagina	**e.** hollow, muscular organ where the embryo develops
	f. female gametes

The Female Reproductive Cycle

Complete each statement by underlining the correct word or phrase in the brackets.

7. Ovulation occurs [at a specific time each month/continuously].

8. Eggs mature inside a fluid-filled chamber called a [fallopian tube/follicle].

9. Estrogen and progesterone levels are [low/high] after ovulation.

10. The uterus [sheds/thickens] its lining if no fertilized egg is received.

■ Section 35-3 *Fertilization and Development* page 728

Fertilization: Fusion of Gametes

Number the following to show the correct sequential order in which sperm travels in the female reproductive system.

_____ **1.** uterus

_____ **2.** cervix

_____ **3.** vagina

_____ **4.** fallopian tubes

The Embryo Enters the Lining of the Uterus

Complete each statement by writing the correct word in each space provided.

5. After the zygote divides into two cells, it is called a(n) _____ .

6. From the ninth week of development until birth the developing embryo is called a(n)

_____ .

7. Doctors can determine if a female is pregnant by testing her urine or blood for the

presence of the _____ produced by the embryo.

The Placenta

Place a check mark next to each accurate statement.

_____ **8.** The mother transfers nutrients and oxygen to the embryo through the placenta.

_____ **9.** The umbilical cord connects the placenta to the embryo.

_____ **10.** The maternal blood and fetal blood mix directly.

Maternal Health

Mark each statement below *T* if it is true and *F* if it is false.

_____ **11.** X rays and ultraviolet rays can cause malformations in a fetus and are, therefore, considered teratogens.

_____ **12.** The child shown in Figure 35-15a has suffered birth defects because of heavy alcohol consumption by the mother after pregnancy.

_____ **13.** Cigarette smoke, caffeine, and other chemicals in the mother's blood stream can cross the placenta.

Growth and Development

On the line at the left, write the letter of the answer that best completes each sentence.

_____ **14.** Pregnancy is divided into
  a. four-month segments called semesters.
  b. three-month segments called quarters.
  c. three-month segments called trimesters.
  d. one-month segments called periods.

_____ **15.** Most of the fetus's development is complete by the end of the
  a. first trimester.
  b. second trimester.
  c. third period.
  d. second semester.

_____ **16.** Infants are considered premature if they are born before the
   a. 37th week of pregnancy.
   b. 38th week of pregnancy.
   c. 39th week of pregnancy
   d. 40th week of pregnancy.

Birth

Read each question, and write your answer in the space provided.

17. Describe what is happening in Figure 35-18a. _____

_____

_____

18. What is taking place in Figure 35-18b? _____

_____

_____

Section 35-4 *Sexually Transmitted Diseases*

page 737

Sexually Transmitted Viral Diseases

Complete each statement by underlining the correct word or phrase in the brackets.

1. AIDS can be contracted through [sexual intercourse/kissing] or unsterilized [toilet seats/needles].

2. A mother [can/cannot] transmit HIV to her infant through breast milk.

3. Hepatitis B infection is caused by [HBV/HIV].

4. Antiviral drugs [can/cannot] eliminate the virus that causes genital herpes from the body.

5. According to Figure 35-20, today's fastest growing STD is [HBV/HIV].

(continued on the next page)

Sexually Transmitted Bacterial Diseases

Compare sexually transmitted bacterial diseases by completing the following table.

	Disease	Cause	Symptoms	Cure	Possible results
6.	Syphilis				
7.	Gonorrhea				
8.	Chlamydia				
9.	Pelvic inflammatory disease				

Answer Key

1 The Science of Biology
pp. 1–5

SECTION 1-1

1. Science is a way of investigating the world, of observing nature in order to form general rules about what causes things to happen.
2. Biology is the study of living things.
3. **a.** T
   b. F
   c. F
   d. T
4. cystic fibrosis, muscular dystrophy
5. genetic engineering
6. Energy use is increasing, more resources are being consumed, and more waste is being produced.
7. Researchers are working to increase the amount of food that can be produced without the use of large amounts of fertilizer, pesticides, and energy-consuming equipment. Scientists are also using genetic engineering to increase the productivity of crop plants by increasing resistance to pests, increasing growth rates, or improving nutritional quality.
8. acid rain, ozone depletion, and rising temperature of the Earth's atmosphere
9. To produce lumber and to make cattle pastures
10. two-thirds
11. Possible responses include:
    Should I smoke?
    How can I avoid developing heart disease?
    What are the risks in taking drugs?
    Will I get cancer?

SECTION 1-2

1. Patients in open wards were more likely to develop the disease than patients in closed wards.
2. malaria
3. *Anopheles* mosquitoes
4. hypothesis
5. predictions
6. *Plasmodium*
7. The parasites traveled from the mosquito's stomach to its salivary glands. The parasites were then transferred to the next person bitten by the mosquito.
8. Ross checked mosquitoes that had bitten only humans who did not have malaria. When he examined these mosquitoes, he found no parasites.
9. control
10. theory
11. observations, control

12. F
13. T
14. T
15. T
16. They use insight and imagination, hunches, or educated guesses. Successful scientists design experiments with a good idea of what they are going to find.

SECTION 1-3

1. T
2. F
3. T
4. T
5. T
6. F
7. T
8. F
9. F
10. T
11. T
12. F
13. evolution
14. survive, reproduce
15. species
16. Ecology is the study of the way living things interact with each other and with the nonliving part of their environment.
17. They may be interrupted when the environment is polluted and individual species become extinct.
18. atoms
19. sun
20. energy

2 Discovering Life pp. 6–9

SECTION 2-1

1. b
2. d
3. a
4. c
5. cellular slime mold
6. cells
7. metabolism
8. homeostasis
9. reproduce, heredity

SECTION 2-2

1. atom
2. elements
3. electrons
4. neutrons and protons
5. energy levels
6. F
7. F
8. T
9. T

10. F
11. T
12. An ionic bond is the force of attraction between oppositely charged (positive and negative) ions. A covalent bond is the force holding atoms together that results from the sharing of electrons.
13. A molecule is a group of atoms held together by covalent bonds, perhaps consisting of different types of atoms. An element is a substance composed of only one type of atom.
14. A single bond is a covalent bond in which two atoms share a set of (two) electrons. A double bond is a covalent bond in which two atoms share two sets of (four) electrons. A triple bond is a covalent bond in which two atoms share three sets of (six) electrons.

SECTION 2-3

1. carbon
2. carbon-carbon
3. carbohydrates, lipids, proteins, and nucleic acids
4. chain reaction
5. T
6. F
7. T
8. F
9. T
10. oil
11. glycerol
12. unsaturated
13. more likely
14. An organic macromolecule composed of long chains of amino acids
15. Amino acids within a protein may attract or repel each other, and some amino acids are repelled by water. Therefore, each protein tends to form complex shapes, each different from other proteins.
16. They play structural roles, and they act as enzymes.
17. d
18. a
19. e
20. b
21. c

3 Cells pp. 10–14

SECTION 3-1

1. cell
2. cell membrane
3. cells
4. F
5. T

6. T
7. water
8. oxygen
9. polar
10. hydrogen
11. The oil is made up of nonpolar lipid molecules and is not attracted to the polar water molecules. When the water molecules are attracted to one another, the oil is pushed away.
12. They have no negative and positive poles.

SECTION 3-2

1. membrane
2. head, tails
3. lipid
4. Polar molecules cannot cross the barrier formed by the lipid bilayer. In addition, the lipid bilayer is fluid, allowing the phospholipid and protein molecules to move from one region of the cell membrane to another.
5. By means of protein passageways through the lipid bilayer
6. Proteins
7. channel
8. receptor
9. Markers
10. F
11. T
12. T

SECTION 3-3

1. eukaryotes, prokaryotes
2. nucleus
3. Bacteria
4. cell membranes, cytoplasm, ribosomes
5. T
6. F
7. F
8. T
9. d
10. a
11. e
12. c
13. b
14. **a.** both
   b. both
   c. both
   d. plants
   e. both
   f. plants
   g. plants
15. Possible responses include: many mitochondria and bacteria are similar in size; mitochondria and some bacteria have a set of double membranes; mitochondria have their own ribosomes and chromosomes, and the ribosomes in mitochondria are

structurally similar to those found in prokaryotes; some organelles divide in a manner similar to bacteria.

16. T
17. F
18. T
19. F

4 The Living Cell pp. 15–17

SECTION 4-1

1. F
2. T
3. receptor proteins
4. membrane
5. hormones
6. glucagon
7. thyroxin, estrogen, testosterone
8. electrical
9. voltage-sensitive
10. opens
11. cell surface markers
12. identical twins

SECTION 4-2

1. b
2. c
3. a
4. During facilitated diffusion a molecule or ion that fits may pass through a selective transport channel in the direction of least concentration. Active transport takes place through a selective transport pump. A cell uses energy to change the shape of the pump in order to move a substance through it to an area of higher concentration.
5. Sodium-potassium pumps help restore the balance required for nerve cells to function properly. In addition, these pumps help move sugars and amino acids into our cells. In humans, proton pumps transform energy from the food we eat into energy that the body can use.
6. The protein channels in the cells of cystic fibrosis patients are defective and cannot export chloride ions, causing high levels of salt in the cells. This salt imbalance causes thick levels of mucus to build up in the patients' lungs, pancreas, and liver.
7. Endocytosis is the process of bringing particles into the cell, and exocytosis is the process of expelling particles from the cell.

5 Energy and Life pp. 18–21

SECTION 5-1

1. c
2. d
3. f
4. a
5. b
6. e
7. carbonic anhydrase
8. substrate
9. active sites
10. acidity

SECTION 5-2

1. T
2. F
3. F
4. T
5. photosynthesis
6. cellular respiration

SECTION 5-3

1. T
2. F
3. visible
4. Retinal
5. chlorophyll
6. thylakoid
7. water
8. 2
9. 4
10. 1
11. 3
12. 5
13. $6CO_2 + 6H_2O$ — (light) $\rightarrow C_2H_{12}O_6 + 6O_2$
14. The Calvin cycle uses carbon atoms from carbon dioxide in the air, energy from ATP, and electrons from hydrogen atoms from NADPH to produce a series of organic compounds. Some of these organic molecules are used to make sugars and other substances needed for energy and growth. Other organic molecules return to the beginning of the cycle, enabling the cycle to continue.

SECTION 5-4

1. glycolysis
2. Oxidative respiration
3. pyruvic acid
4. fermentation, oxidative respiration
5. ethyl alcohol
6. lactic acid
7. T
8. F
9. T
10. F
11. T
12. Cellular respiration is regulated by a process called feedback inhibition. Key reactions early in glycolysis and the Krebs cycle are catalyzed by enzymes that have a second regulatory active site. When the body's cells have enough ATP, the ATP molecules bind to this second active site

on the enzyme, causing the enzyme to change its shape. When the enzyme's shape changes, it can no longer accommodate its substrate. Thus the enzyme becomes inactive, and no more ATP is produced.

6 Cell Reproduction
pp. 22–24

SECTION 6-1

1. chromosome
2. chromatin
3. 5
4. T
5. F
6. T
7. T
8. F

SECTION 6-2

1. genetic information or DNA
2. bacteria
3. During eukaryotic cell division each chromosome must be copied exactly, then the chromosomes must be sorted out so that each new cell gets a complete set of chromosomes in its nucleus. During prokaryotic cell division, the cell's single circle of DNA, or chromosome, makes a copy of itself so that each new cell gets one of the two resulting circles of DNA.
4. Prophase, metaphase, anaphase, and telophase
5. Two cells with the same genetic information
6. 3
7. 5
8. 1
9. 4
10. 2
11. cyclins
12. tumor
13. the cell cycle
14. cancer

SECTION 6-3

1. Gametes
2. meiosis
3. half
4. T
5. T
6. F
7. F

7 Genetics and Inheritance pp. 25–28

SECTION 7-1

1. T

2. F
3. T
4. F
5. white
6. purple
7. 3:1
8. a
9. c
10. f
11. b
12. g
13. d
14. e
15. Punnett square
16. genotype
17. probability
18. *WW, Ww,* or *ww*
19. 3:1
20. segregation
21. independent assortment

SECTION 7-2

1. monohybrid
2. One phenotype, purple, and one genotype, *Ww*
3. Y = yellow, y = green, R = round, and r = wrinkled
4. 9 yellow round seeds : 3 yellow wrinkled seeds : 3 green round seeds : 1 green wrinkled seed
5. incomplete dominance
6. codominance
7. $I_A I_A$ or $I_A i$
8. polygenic

SECTION 7-3

1. T
2. F
3. T
4. Cystic fibrosis
5. malaria
6. sons
7. Down syndrome
8. They may wish to determine if they are at risk for genetic disorders. A family pedigree can show inheritance patterns over several generations and help the couple determine their chances for carrying genetic disorders. With this information, they can determine whether or not they wish to have children.
9. Phenylketonuria
10. defective, healthy

8 How Genes Work pp. 29–31

SECTION 8-1

1. virulent
2. transformation
3. DNA

4. Hershey, Chase
5. b
6. a
7. d
8. c
9. F
10. T
11. F
12. T

SECTION 8-2

1. ribonucleic acid or RNA
2. gene expression
3. transcription
4. translation
5. DNA
6. uracil
7. transcription
8. RNA
9. codon
10. UUU
11. 64
12. bacteria
13. 2
14. 3
15. 4
16. 1

SECTION 8-3

1. T
2. F
3. T
4. F
5. A eukaryotic gene contains a series of sequences called exons and introns. Exons are the portions of a gene that get translated into proteins. Exons are interrupted by noncoding portions of DNA called introns.
6. Genes known as transposons can jump to new locations on genes, sometimes creating mutations.

9 Gene Technology
pp. 32–35

SECTION 9-1

1. DNA
2. bacteria
3. genetic
4. T
5. F
6. T
7. F
8. T
9. F
10. Two concerns are the effects of genetically engineered organisms on the environment and the possibility of creating new microorganisms for biological warfare.

SECTION 9-2

1. T
2. F
3. T
4. If farmers plant crops that are resistant to glyphosate, they can treat their fields with this powerful herbicide and all growing plants will die except for their crops. Other advantages are that glyphosate breaks down quickly, it is not toxic to humans, and it prevents erosion since intense cultivation to remove weeds is not necessary.
5. Since plants cannot obtain nitrogen from the air, some plants obtain this essential element through nitrogen fixation. In this process bacteria living in the roots of plants such as soybeans, peanuts, and clover convert nitrogen gas from the atmosphere into a form that plants can use.
6. cotton
7. tomato
8. F
9. T
10. T

SECTION 9-3

1. diabetes
2. insulin
3. 3
4. 4
5. 2
6. 1
7. (a) The defective gene that caused the disease was difficult to identify and isolate. (b) It was hard to transfer a healthy copy of such a gene into the cells of body tissues. (c) There was no way to keep the altered cells or their offspring alive in the body for a long time.
8. In 1993, doctors inserted a missing gene into blood cells of three infants' umbilical cords. They then injected these cells into the infants' bloodstreams. In 1995, these three children were producing the blood factor that would likely have been missing had the gene therapy not taken place.
9. They have developed a way to add the gene encoding TNF to a type of white blood cell that is effective in locating cancer cells.
10. c
11. a
12. d
13. b

Ch. 10 Answer Key

10 Evolution and Natural Selection pp. 36–39

SECTION 10-1
1. T
2. F
3. T
4. Galapagos
5. evolution
6. Darwin's
7. more

SECTION 10-2
1. fossils
2. sediment
3. F
4. T
5. F
6. T
7. d
8. a
9. b
10. c

SECTION 10-3
1. 4
2. 1
3. 5
4. 3
5. 2
6. Step 1: More dark moths survived when equal numbers of light and dark moths were released in a forest with dark tree trunks. Two-thirds of the moths recaptured in this step were dark moths.
7. Step 2: More light moths survived when equal numbers of light and dark moths were released in a forest with light tree trunks. Two-thirds of the moths recaptured in this step were light moths.
8. Step 3: Cameras showed that birds ate more light moths in the forest with dark tree trunks. In the forest with light tree trunks, the birds were more likely to eat dark moths.
9. sickle cell anemia
10. malaria
11. balancing
12. directional
13. F
14. T
15. T
16. Gradualism is the hypothesis that evolution occurs at a slow, constant rate. Punctuated equilibria is the hypothesis that evolution occurs at an irregular rate.

11 History of Life on Earth pp. 40–43

SECTION 11-1
1. extraterrestrial
2. creationists
3. nonliving or inanimate
4. c
5. b
6. a
7. T
8. T
9. F
10. F

SECTION 11-2
1. eubacteria
2. archaebacteria
3. cyanobacteria
4. a, d
5. a, d
6. b, e
7. c, e
8. c, e
9. c, e
10. the Cambrian period
11. phyla
12. a rich collection of extinct Cambrian fossils unlike any animals living today
13. at least five

SECTION 11-3
1. T
2. F
3. T
4. mycorrhizae
5. mutualism
6. arthropods
7. T
8. F
9. T
10. largest
11. mutualism

SECTION 11-4
1. c
2. e
3. h
4. a
5. i
6. b
7. j
8. d
9. g
10. k
11. f

Ch. 12 Answer Key

12 Human Evolution pp. 44–47

SECTION 12-1
1. F
2. T
3. F
4. T
5. T
6. F
7. eye sockets
8. anthropoids
9. brains
10. apes, humans
11. gibbons
12. bonobos and chimpanzees
13. The nucleotide sequences of human and chimpanzee genes differ only by 1.6 percent, indicating a very close evolutionary relationship.

SECTION 12-2
1. T
2. F
3. T
4. F
5. Africa
6. bipedal or upright
7. larger, smaller
8. bipedal, brain larger than ape; 1924; South Africa
9. massive teeth and jaws; 1938; South Africa
10. bony ridge on crest of head, powerful jaw muscles; 1959; Tanzania
11. small stature, bipedal, brain the size of chimpanzee; 1974; Ethiopia

SECTION 12-3
1. *Homo habilis, Homo erectus, Homo sapiens*
2. Most researchers think *A. afarensis* is the ancestor of all humans. However, in one theory, *Homo sapiens* then descended from *H. habilis* and *H. erectus*. In the other theory, the line of descent passes from *H. africanus, H. habilis,* and *H. erectus* to *Homo sapiens.*
3. fossils, tools
4. two
5. large, small, tools
6. 225
7. Africa
8. Asia, Europe
9. *Homo sapiens* first evolved in Africa and then spread to the rest of the world.
10. They lived in huts and caves, had diverse tools, took care of their injured, and buried their dead.
11. at least 100,000 years ago
12. increase in brain size
13. Humans have found ways to mold the environment to meet their needs.

13 Animal Behavior pp. 48–51

SECTION 13-1
1. behavior
2. mechanisms
3. evolution
4. T
5. F
6. T
7. F
8. There is a genetic basis for the nest-building behavior in these birds.
9. They tried unsuccessfully to carry the nesting material in their feathers, and usually ended up carrying it in their beaks.
10. After four generations of breeding long-singing males and short-singing males, two distinct groups were produced that differed in the amount of time they chirped. The difference existed because only the extreme individuals were allowed to mate, thus genetically transmitting their chirping length behaviors.
11. b
12. c
13. a
14. fewer
15. hereditary
16. no

SECTION 13-2
1. Answers may vary. Parental care ensures the survival of young; courtship behavior attracts mates; defensive behavior protects individuals from predators.
2. T
3. F
4. F
5. T
6. chemical
7. sexual selection
8. reproductive tract
9. Answers may vary but should indicate that altruism is self sacrificing behavior.
10. Answers may vary. One example is worker bees that care for and defend the queen or that sting an attacker even though this will result in their death. Jays that remain in the nest with parents to help raise other offspring are another example.
11. An individual may pass on its own genes by helping its relatives to reproduce rather than by reproducing itself.

14 Ecosystems pp. 52–55

SECTION 14-1
1. T
2. F
3. b
4. d
5. a
6. e
7. c
8. thousands of interacting species
9. cannot
10. only as accurately as the information used in building the model
11. consumer; herbivore; 2
12. consumer; omnivore; 3
13. producer; not applicable; 1
14. F
15. T
16. T
17. F

SECTION 14-2
1. Does rainwater remove nutrients from ecosystems?
2. They found that cutting the trees and vegetation greatly reduced the ecosystem's ability to retain nutrients.
3. No. Nitrogen in the atmosphere is made of two nitrogen atoms bound together by a strong bond that is difficult to break.
4. nitrogen fixation
5. T
6. F
7. T
8. atmosphere
9. carbon dioxide
10. greenhouse effect

SECTION 14-3
1. plankton
2. shallow, surface, light
3. e
4. d
5. a
6. c
7. g
8. f
9. b
10. open ocean surface
11. shallow ocean waters
12. deep ocean waters

15 How Ecosytems Change pp. 56–59

SECTION 15-1
1. interactions with the physical environment and interactions with other living members of the ecosystem

2. Coevolution is the process that occurs when two or more species evolve in response to each other. For example, many plants evolve tough leaves to protect against herbivores, while the herbivores evolve flatter, larger teeth to grind the leaves.
3. coevolution
4. pollinators
5. rotting flesh
6. T
7. F
8. F
9. T
10. mutualism
11. commensalism
12. parasitism
13. mutualism
14. commensalism
15. parasitism

SECTION 15-2
1. F
2. T
3. T
4. competing
5. realized
6. death
7. decrease
8. 2
9. 4
10. 1
11. 3
12. F
13. T
14. T
15. F
16. a. T
    b. F
    c. T
    d. F
    e. T
    f. F
17. Disrupting the physical habitat is one of the principal ways humans disrupt ecosystems. An example of this is cutting forests. Another way humans disrupt ecosystems is by decreasing species diversity. This happens when forests are converted to farmland. The destruction of interactions among species is the other principal way humans disrupt ecosystems. The removal of predators from ecosystems is an example of this.

16 The Fragile Earth pp. 60–64

SECTION 16-1
1. T
2. F

3. T
4. F
5. Eighty
6. *Exxon Valdez*
7. acid rain
8. H+ or hydrogen
9. a
10. c
11. d
12. Burning fossil fuels
13. Carbon dioxide and other greenhouse gases such as CFCs and methane in the atmosphere hold the sun's heat in Earth's atmosphere, raising the temperature at Earth's surface.
14. Increased levels of carbon dioxide in the atmosphere are causing increases in global temperatures.
15. 1.5 to 4.5 degrees Centigrade

SECTION 16-2
1. T
2. T
3. F
4. T
5. F
6. T
7. F
8. topsoil
9. ground water
10. half or 50 percent
11. extinct
12. cancer
13. c
14. a
15. d

SECTION 16-3
1. solvable
2. Many
3. have been
4. five times
5. Answers will vary. See Table 16-1.
6. Answers will vary. See Table 16-1.
7. Answers will vary. See Table 16-1.
8. Answers will vary. See Table 16-1.
9. Answers will vary. See Table 16-1.
10. In the United States, 5 percent of the world's population uses 25 percent of the world's energy.
11. Answers will vary. Examples include turning off lights and appliances when leaving the room, using fluorescent rather than incandescent light bulbs, setting hot-water temperatures lower, reducing the use of automobiles, and making sure car tires are inflated properly and that cars receive regular maintenance.
12. Answers will vary. Examples include recycling, disposing of toxic and hazardous wastes properly, turning off faucets and taps when not in use, reducing the amount of water used to flush a toilet, and installing low-flow shower heads.
13. b

17 Classifying Living Things
pp. 65–68

SECTION 17-1
1. scientific name
2. Homo sapiens
3. organism or living thing
4. T
5. T
6. F
7. F
8. T
9. T

SECTION 17-2
1. b
2. e
3. a
4. f
5. d
6. c
7. common ancestor
8. convergent evolution
9. analogous structures
10. C
11. P
12. C
13. P
14. Techniques of molecular biology have enabled taxonomists to compare the DNA sequences of different organisms.
15. As time passes, more mutations tend to accumulate in the DNA of a particular species.
16. The more similar the DNA sequences of two species, the more recently their common ancestor must have lived, and the more closely they are related.
17. T
18. T
19. F
20. T
21. F

SECTION 17-3
1. Plantae
2. six
3. animal
4. T
5. T
6. F
7. F
8. T
9. T
10. F

18 Bacteria and Viruses
pp. 69–72

SECTION 18-1
1. almost everywhere
2. three
3. prokaryotes
4. Gram-positive
5. unaffected
6. by splitting in two
7. Bacteria are able to use a wide variety of food and energy sources, including sunlight, living and dead tissue, and inorganic molecules.
8. They capture energy from the sun.
9. inorganic molecules
10. dead animals; animal wastes; live animals; dead plants; fallen leaves, fruit, and branches

SECTION 18-2
1. decomposers
2. ammonia
3. bacteria
4. T
5. F
6. T
7. F
8. T
9. T
10. F
11. c
12. a
13. d
14. b

SECTION 18-3
1. T
2. F
3. T
4. F
5. 3
6. 2
7. 5
8. 1
9. 4
10. cannot
11. viruses
12. eradicated
13. unidentified animals
14. mutate frequently
15. yellow fever and the Ebola virus

19 Protists pp. 73–75

SECTION 19-1
1. F
2. T
3. T
4. F
5. F
6. T
7. T
8. F
9. three
10. multicellular eukaryotic kingdoms

SECTION 19-2
1. F
2. T
3. T
4. e
5. d
6. b
7. c
8. a
9. f
10. heterotrophic
11. differ from
12. animals
13. a separate kingdom of their own

SECTION 19-3
1. insects, contaminated water
2. one million
3. *Plasmodium*
4. quinine
5. insecticides
6. evolutionary
7. immunity
8. industrialized
9. toxoplasmosis
10. sleeping sickness, Chagas' disease

20 Fungi and Plants pp. 76–80

SECTION 20-1
1. plant
2. chlorophyll
3. hyphae
4. chitin
5. digest, absorb
6. spore
7. c
8. a
9. b
10. a
11. d
12. c
13. Fungi on or within the roots of certain plants provide the plants with water and

minerals. The plants provide the fungi with photosynthetically produced food.

14. The destruction of mycorrhizae by acid rain causes the forest trees to be unable to absorb minerals from the soil.

15. The fungus shields the alga from excessive sunlight and retains the water that the alga needs for photosynthesis, while the alga secretes carbohydrates that the fungus needs for food.

16. *Saccharomyces cerevisiae*

17. *Penicillium notatum*

18. Organ transplants. Cyclosporine suppresses the immune system's response to transplanted organs.

Section 20-2

1. photosynthetic; cellulose
2. on land
3. vascular
4. cuticle
5. gametophyte, sporophyte
6. T
7. F
8. T
9. F
10. T
11. larger than
12. seedless
13. sporophyte

Section 20-3

1. woody tissue and complex vascular tissue
2. pollen, cones or flowers, and seeds
3. embryo, cotyledons, and outer seed coat
4. hooks, feathery projections, and winglike projections
5. T
6. F
7. T
8. F
9. T
10. flowers
11. insects, birds, animals
12. 235,000
13. Brassicaceae

21 Plant Form and Function
pp. 81–83

Section 21-1

1. T
2. F
3. T
4. shoot
5. leaves
6. chloroplasts
7. d
8. f
9. a

10. e
11. c
12. b
13. T
14. F

Section 21-2

1. osmosis
2. capillary action
3. transpiration
4. stomata
5. translocation
6. T
7. F
8. T
9. T
10. F
11. F

Section 21-3

1. stamen
2. pistil
3. d
4. c
5. b
6. a
7. T
8. F
9. T
10. 3
11. 1
12. 4
13. 2

22 Plants in Our Lives
pp. 84–87

Section 22-1

1. wheat, rice, and corn
2. Answers will vary. Possible responses include breads, pasta, tabblouleh, pilavi, and kisir.
3. Whole wheat flour contains the most protein and the most fiber.
4. livestock
5. d
6. c
7. a
8. e
9. b
10. T
11. F
12. T
13. T
14. F
15. T
16. cotton
17. pine trees
18. latex

SECTION 22-2

1. c
2. f
3. e
4. g
5. a
6. i
7. b
8. h
9. d
10. T
11. T
12. long
13. pruned
14. caterpillars and pill bugs
15. planting root cuttings, dividing bulbs, rooting leaf cuttings, burying part of the stem in soil, and placing a small piece of plant tissue in nutrient-rich soil
16. 4
17. 3
18. 1
19. 5
20. 2

23 The Animal Body pp. 88–91

SECTION 23-1

1. body plan
2. protist
3. F
4. T
5. T
6. b
7. e
8. d
9. a
10. c
11. F
12. T
13. T

SECTION 23-2

1. e
2. c
3. d
4. f
5. b
6. a
7. T
8. T
9. F
10. coelom
11. digestive
12. circulatory system
13. coelom, circulatory system

SECTION 23-3

1. segmentation
2. tube within a tube
3. jointed appendages; exoskeleton
4. chitin
5. T
6. F
7. T
8. deuterostome development; endoskeleton
9. five-part
10. T
11. F
12. T

24 Adaptation to Land
pp. 92–95

SECTION 24-1

1. T
2. F
3. skeleton
4. limb structure
5. reptile
6. their lateral line system
7. through the ear where the eardrum and three small bones amplify sounds and transmit them to the inner ear where the sounds are detected
8. membranes on their front pair of legs

SECTION 24-2

1. skin
2. exoskeleton
3. mucus
4. scales
5. F
6. F
7. T
8. T
9. inhalation, capillaries
10. vertebrates
11. rapidly
12. T
13. T
14. F
15. b
16. c
17. e
18. a
19. d

SECTION 24-3

1. c
2. d
3. e
4. b
5. a
6. f
7. Once the egg has been laid, the parent

cannot provide further nourishment to the offspring until it hatches. In addition, the egg is exposed to environmental hazards such as predators and extreme weather.
8. platypuses and echidnas
9. inside its mother
10. The mammal embryo is receiving nourishment and discarding wastes through the mother's placenta.

25 Animal Diversity pp. 96–99

SECTION 25-1
1. F
2. T
3. T
4. F
5. b
6. a
7. c
8. d
9. 1
10. 2
11. 4
12. 5
13. 3
14. 6
15. T
16. F
17. F
18. T
19. T

SECTION 25-2
1. Nephridia
2. Gastropoda
3. radula
4. bivalve
5. nervous
6. T
7. F
8. T
9. F
10. 1 million
11. segmented, exoskeleton
12. lobsters, crabs, shrimp

SECTION 25-3
1. five-part
2. water vascular system
3. notochord, dorsal nerve cord, post-anal tail, pharyngeal slits
4. tunic
5. pharyngeal slits
6. Branchiostoma
7. increased complexity of many organ systems and internal skeleton
8. vertebral column

26 Arthropods pp. 100–102

SECTION 26-1
1. c
2. a
3. b
4. land
5. book lungs
6. spinnerets
7. web
8. courtship
9. T
10. F
11. T
12. F
13. T

SECTION 26-2
1. e
2. c
3. d
4. b
5. a
6. g
7. h
8. f
9. T
10. F
11. T

SECTION 26-3
1. eight
2. less
3. efficient excretory organs
4. gills
5. F
6. F
7. T
8. T
9. F
10. T

27 Fishes and Amphibians pp. 103–105

SECTION 27-1
1. b
2. a
3. c
4. d
5. jaws
6. food
7. acanthodians
8. agnathans
9. T
10. T
11. F
12. F

SECTION 27-2

1. T
2. T
3. F
4. T
5. F
6. c
7. e
8. a
9. f
10. g
11. b
12. h
13. d

SECTION 27-3

1. walk on land
2. caecilians
3. lungs
4. moist
5. circulatory
6. 3
7. 4
8. 1
9. 2
10. none; short or absent; tropics; caecilians
11. hind specialized for jumping; none; worldwide; frogs, toads
12. set at right angles to body; distinct; worldwide except Australia; salamanders, newts

28 Reptiles, Birds, and Mammals pp. 106–111

SECTION 28-1

1. c
2. b
3. a
4. d
5. aquatic; resembled dolphins, fed on fish
6. aquatic; barrel-shaped bodies with paddle-like fins, fed on fish
7. land; herbivorous
8. land; carnivorous
9. land/air; flew like birds, ranged from sparrow-sized to 11 m (35 ft) wingspan
10. T
11. T
12. F
13. T
14. T
15. F
16. F
17. F

SECTION 28-2

1. They lay amniotic eggs and have scales on their feet and lower legs.
2. Most of a bird's feathers are contour feathers. They cover the bird's body and give its wings and tail their shape and provide insulation. Down feathers grow underneath or among the contour feathers and they are specialized to provide insulation.
3. A bird's flight muscles are secured to its large sternum.
4. Birds' hearts have four chambers with separate circulatory loops to the lungs and to the body. This system allows a bird to rapidly get oxygen-rich blood to the tissues where it is needed during flight.
5. T
6. F
7. T
8. F
9. Apodiformes
10. parrot or cockatoo
11. eagle, hawk, falcon, or vulture
12. eyes
13. penguin

SECTION 28-3

1. The early mammals had a jaw consisting of only one bone, while the therapsids had a jaw made up of several bones.
2. Mammals have diversified and taken over the ecological roles formerly occupied by dinosaurs.
3. The cooling of the world's climate and the subsequent decline in rain forest coverage worldwide has led to a decline in the number of mammalian species.
4. T
5. F
6. T
7. Monotremes
8. platypus, echidna
9. fur, mammary
10. T
11. T
12. F
13. c
14. f
15. i
16. a
17. g
18. k
19. j
20. b
21. e
22. l
23. h
24. d

SECTION 28-4

1. guard hair and underhair
2. insulate it against heat loss
3. sensory structures
4. T

5. F

6. T

7. The coyote has long canines to bite and hold prey along with sharp molars and premolars to cut off chunks of flesh. A deer clips off mouthfuls of plants with its flat incisors and grinds plants with its large, flat molars and premolars.

8. Bacteria and protists that live in the first chamber of the herbivores' four-chambered stomachs

9. When eaten in similar quantities, plants contain less nutrition than flesh.

10. T

11. F

12. T

29 The Human Body
pp. 112–116

SECTION 29-1

1. f

2. c

3. e

4. b

5. d

6. a

7. organ

8. skin, tissue

9. organ system

10. integumentary, muscular, skeletal

SECTION 29-2

1. e

2. b

3. d

4. f

5. c

6. a

7. F

8. T

9. T

10. T

11. F

12. oil, bacteria, cellular debris

13. more rapid, skin cancer

14. malignant melanomas

15. p53 gene

SECTION 29-3

1. c

2. d

3. b

4. a

5. c

6. e

7. f

8. h

9. b

10. a

11. i

12. d

13. g

14. c

SECTION 29-4

1. attached to bones of skeleton; moves bones of skeleton

2. internal organs; makes internal organs work

3. heart; causes heart to contract

4. T

5. T

6. F

7. T

8. F

9. F

10. Aerobic exercise takes place at a slow to moderate pace, enabling the lungs and heart to convey the oxygen to the muscles at the same rate at which the muscle cells are using it. Anaerobic exercise takes place in intense, short bursts, without allowing time for ample oxygen intake and, thus, resulting in a shortage of oxygen in the muscle cells.

11. Fast-twitch muscle cells are activated when speed over a short period is required. Slow-twitch muscles are called into action when endurance is required.

12. Anabolic steroids are powerful, synthetic compounds that are used to increase muscle size but have dangerous consequences, including incomplete bone growth and the development of sexual characteristics of the opposite sex. Testosterone is the natural male sex hormone that chemical steroids approximate.

13. Muscle strain may occur when a muscle is overstretched or torn due to overuse or inadequate warm-ups. Tendinitis is an inflammation of a tendon due to overuse or stress on the tendon.

30 The Nervous System
pp. 117–120

SECTION 30-1

1. b

2. d

3. a

4. c

5. resting potential

6. action potential

7. nerve impulse

8. T

9. T

10. F

11. F

12. F
13. T
14. F
15. nervous
16. serotonin
17. receptors
18. dopamine
19. cell membrane

Section 30-2

1. T
2. F
3. F
4. b
5. d
6. a
7. c
8. Learning
9. left
10. magnetic resonance imaging
11. PET
12. peripheral nervous system
13. sensory neurons
14. motor neurons
15. reflex
16. autonomic nervous system

Section 30-3

1. Heat receptors and cold receptors respond to temperature changes. Mechanoreceptors respond to pressure.
2. On the tongue
3. The stimulus of light initiates the process of focusing and visual processing.
4. T
5. F
6. T
7. two
8. able
9. about three-quarters
10. are not
11. T
12. T
13. F

31 Hormones pp. 121–124

Section 31-1

1. T
2. F
3. T
4. F
5. e
6. d
7. a
8. c
9. b
10. feedback
11. Positive, negative

12. body temperature
13. calcitonin, parathyroid hormone

Section 31-2

1. A hormone is binding to a receptor protein in the cell membrane. The receptor protein then changes shape, causing changes to take place in the cell.
2. Receptor proteins may be located in the cell membrane, inside the cytoplasm, or in the nucleus of the cell.
3. T
4. T
5. F
6. F
7. Peptide
8. cell membrane
9. cyclic AMP
10. many

Section 31-3

1. adrenal glands
2. adrenal medulla, adrenal cortex
3. epinephrine or adrenaline, norepinephrine or noradrenalin
4. homeostasis
5. F
6. T
7. T
8. F
9. b
10. a
11. f
12. e
13. d
14. c
15. F
16. T
17. T
18. F

32 Circulation and Respiration pp. 125–129

Section 32-1

1. F
2. T
3. T
4. e
5. a
6. b
7. c
8. d
9. carry blood away from the heart; elastic, muscular walls
10. return blood to the heart; slightly elastic, valves prevent backward flow
11. carry blood to and from tissues; walls one cell thick connect veins and arteries

Ch. 32 Answer Key, *continued*

12. F
13. T
14. F
15. Lymph nodes are located in groups in various places along the lymphatic vessels. These vessels, located throughout the body, carry excess fluids, proteins, and other nutrients back to the blood.
16. Lymph nodes filter foreign matter from the fluid and prevent harmful agents from entering the bloodstream.

SECTION 32-2

1. d
2. i
3. e
4. h
5. a
6. g
7. b
8. f
9. c
10. 7
11. 3
12. 5
13. 6
14. 1
15. 4
16. 2
17. T
18. F
19. T
20. disease
21. homeostasis
22. HDL cholesterol
23. It illustrates atherosclerosis, the partial blockage of an artery due to plaque deposits.
24. Hypertension is the condition that exists when blood pressure against an artery wall remains higher than normal due to blockage in the blood vessels.

SECTION 32-3

1. 3
2. 5
3. 1
4. 6
5. 4
6. 2
7. T
8. F
9. brain and circulatory system
10. greater
11. airways in the lungs become constricted due to sensitivity to certain stimuli
12. lung tissue loses elasticity, reducing efficiency of gas exchange
13. growth of cancerous cells in lung tissue, almost always caused by smoking

33 The Immune System
pp. 130–133

SECTION 33-1

1. body cells
2. mucous membranes
3. enzymes
4. b
5. d
6. a
7. e
8. f
9. c
10. T
11. F
12. T
13. F

SECTION 33-2

1. It is most likely to enter by way of small water droplets expelled from coughs or sneezes of an infected person that penetrate the cells of the mucous membranes lining the respiratory tract.
2. It is the immune system's principal assault on a specific pathogen.
3. T
4. F
5. T
6. macrophages
7. plasma cells
8. bind to
9. T
10. F
11. T
12. F

SECTION 33-3

1. non-pathogenic
2. autoimmune diseases
3. multiple sclerosis
4. T
5. F
6. T
7. macrophages, helper T
8. blood cells, body fluids
9. one-in-three
10. F
11. T
12. T

34 Digestion and Excretion
pp. 134–137

SECTION 34-1

1. Too much food (over-consumption) and too little exercise (sedentary lifestyle)
2. Carbohydrates, protein, fats, water, minerals, and vitamins

3. g
4. b
5. a
6. f
7. e
8. d
9. c
10. water
11. minerals
12. Vitamins
13. Afflicted starve themselves; Lack sexual maturity; muscle and bone mass lost; hair and skin become dry; nails get brittle
14. Binge-purge eating cycles; Trauma to lining of mouth, stomach, and esophagus; erosion of tooth enamel; fatigue; possible death by heart failure, ruptured stomach, or kidney failure

SECTION 34-2

1. digestive enzymes
2. esophagus
3. peristalsis
4. pepsin
5. c
6. g
7. b
8. f
9. a
10. e
11. d

SECTION 34-3

1. c
2. a
3. d
4. F
5. T
6. T
7. F
8. kidney stones
9. surgical or medical
10. hemodialysis

35 Reproduction and Development pp. 138–142

SECTION 35-1

1. See Figure 35-1 on page 719 of student edition.
2. 4
3. 2
4. 3
5. 1
6. T
7. F
8. T
9. T
10. F

SECTION 35-2

1. f
2. a
3. b
4. e
5. d
6. c
7. at a specific time each month
8. follicle
9. high
10. sheds

SECTION 35-3

1. 3
2. 2
3. 1
4. 4
5. embryo
6. fetus
7. hormone
8. T
9. T
10. F
11. T
12. F
13. T
14. c
15. b
16. a
17. The cervix is forced open by the baby's head, and the baby is pushed out through the vagina.
18. The fluid, blood, placenta, and umbilical cord are about to be expelled from the uterus, completing the process of childbirth.

SECTION 35-4

1. sexual intercourse, needles
2. can
3. HBV
4. cannot
5. HBV
6. bacterium; chancre on genitals, disappears, then rash, fever, joint pain; antibiotics; damage to nervous system, blood vessels, skin
7. bacterium; painful urination, pus discharge; antibiotics; damage to urethra, vas deferens, fallopian tubes; infertility
8. bacterium *Chlamydia*; painful urination, vaginal discharge; antibiotics; spread to fallopian tubes, infertility
9. untreated gonorrhea or chlamydia; scar tissue closes fallopian tubes; antibiotics; infertility, ectopic pregnancy